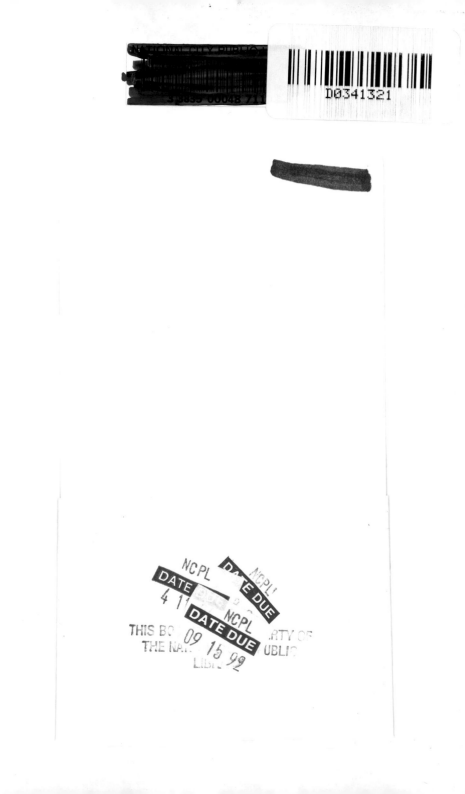

ANIMAL TALK

ANIMAL TALK

Science and the Voices of Nature

EUGENE S. MORTON
JAKE PAGE

RANDOM HOUSE
NEW YORK

Library of Congress Cataloging-in-Publication Data

Morton, Eugene S.
Animal talk: science and the voices of nature/Eugene S. Morton,
Jake Page.—1st ed.
p. cm.
Includes index.
ISBN 0-394-58337-X
1. Animal communication. 2. Sound production by animals.
3. Evolution (Biology) I. Page, Jake. II. Title.
QL776.M66   1992
591.59—dc20   90-52894

Manufactured in the United States of America
24689753
First Edition
Book design by Jo Anne Metsch

To our parents

Acknowledgments

The authors are grateful to Lew Binford, Richard Dawkins, Kim Derrickson, John Eisenberg, Sheri Gish, Judith Hand, Devra Kleiman, Jane Lancaster, Olga Linares, Richard S. Miller, Vickie McDonald, Martin Moynihan, Don Owings, Stan Rand, Eyal Shy, and many other colleagues who listen to animals. The authors are also grateful to Random House editors Sam Vaughan, Becky Saletan, and Jean-Isabel McNutt, who labored with patience, understanding, and skill as readers' advocates to make this elaborate argument in a complex field as clear and as engaging as possible. And, of course, to Letty and Susanne, who helped us play in these fields.

Contents

THE VOICES OF NATURE

Shrieks, howls, grunts, roars.

Burbles and bleats, hums, sweet music, barks, whines, snarls, purring.

Chatters, chirps, moans. Hoots in the night, incessant buzzing, a chorus of croaks. Whistles, hisses, the rattle of drums. Songburst before dawn. Vesper hymns. Silence.

There isn't much silence in nature. The world is full of voices, a cacophony of sound, seeming intention, the array of living things engaged in some kind of communication as they go about their lives. And over it all, the incessant talk of human beings, some of it, almost surely from times immemorial, times that ushered in what we call human nature, devoted to trying to figure out what all those other sounds mean, what the animals are saying.

In the eyes of animals we have always seen a distant mirror of ourselves—so different but clearly with much in common. In the sounds they emit, in their multitudinous voices, we have sought to hear . . . what?

Wisdom. Moral tales.

Warnings: of a dangerous presence nearby of predators, of a change in the weather, of impending earthquakes.

Assurances: that the sun will come up, that we reign supreme, that we are not alone.

Useful analogues: Can we learn peace from dolphins? Can we learn how better to teach our children from teaching chimpanzees to make sentences? We cock our heads to hear, to listen, to understand, what our dog is telling us, our cat, the birds in the yard. Like those patterns we see with our eyes squinched shut, which drift away as we look at them, the meaning of the voices of nature has always seemed to stay beyond our ken.

It is only recently, within less than half a human lifetime, that we have learned the way to really listen to these voices and to understand what they signify. We needed not just hearing aids, but thinking aids, listening devices that, ironically, were first created to pursue war and prevent crime, and we needed to learn to think about these newly and more accurately perceived sounds from the standpoint of what the animals themselves might be hearing. We had to put aside one of our own crowning achievements as a species—our mastery of words and speech—in order to make sense of the other tantalizing voices of nature. We had to listen to creatures that do not, in the sense we use the word, talk. The title of this book is deliberately ironic.

This is by no means to say that most people cannot understand what some animals are trying to get across. It doesn't take much effort to understand that a dog that faces the door and barks in a particular way is drawing attention to the fact that it would rather be on the other side of the door. We train our dogs (well or badly), but it is also clear that, in some transactional sense, they train us. We learn not just what some of their vocalizations mean but their "body language" too, and vice versa. Many people have an intuitive knack for dealing with animals, even quite wild and ferocious ones, or wild and shy ones. Some zoo keepers have an inexplicable ability to make life in captivity sufficiently salubrious that an exotic animal pair will breed. Others are less successful, and no one has worked out exactly what the difference is. Certain people are artists that way, intuitively able to put themselves inside the skin (or the skull) of the animal. If they talk about how they do it, it is generally pretty unconvincing, perhaps because their talent lies outside, beyond—before?—our capacity to talk. It is probably a matter of tapping some preverbal aspect of our being, a kind of mind reading in the sense of empathy, and long-term and mostly unconscious observation through all of the senses. Such a person "knows" what the animal is up to, often without necessarily knowing how he or she knows.

But this is a book of—and about—science, which is a different way of knowing, a formal and even skeptical way of knowing. In science, an anecdotal account is not considered much of a contribution to knowledge. It is a one-time event, hard to verify, hard to put to a test. In fact, science can be thought of not as knowledge at all, but as a way of dispelling ignorance about the world. It is a way of asking questions. To ask what an animal is saying may be the wrong question. It implies that all one really needs to arrive at a reasonable translation of their language into ours is an animal-human dictionary. It implies that there is a wonderful array out there of far and near approaches to human language, varying degrees of pidgin Human.

A great deal of this book is devoted to finding another way of phrasing the questions about the significance of animal utterances, the better to understand them. To accomplish such a task, this book will take the reader on some unexpected paths, some of them quite labyrinthine. It is a quest—among other things—for origins. For example, how did the elaborately melodious songs of birds come about? Without knowing this, it would seem difficult to probe their significance. Necessarily, this book also probes the origins of what has been a certain amount of scientific confusion about the nature of animal communication, which is not to say that all our ignorance about the voices of nature has been dispelled. Not by a long shot. But the barriers to our understanding, almost totally opaque merely a few decades ago, are much less so today.

For example, it is only since World War II that technological devices have permitted us to hear a broader range of voices than we had previously imagined. We have now heard the sounds of creatures that live in the sea—from the staticlike racket of shrimp to the beguiling songs of whales; we have also tuned in on the sounds around us that we cannot hear because they are too high or too low for our ears, the high-pitched calls of bats and rats, the recently perceived sounds below the unaided human hearing threshold that guide the lives of elephants. This same equipment—high-capacity microphones, high-fidelity tapes and speakers, machines that produce pictures of sounds—has also permitted us to listen with greater clarity to the signals of the more familiar virtuosos—bees, birds, and our close relatives, the monkeys and apes. So sudden a burst of information required a framework, a theory, an organized way of looking at all these voices and what they might signify. Several lines of inquiry, several means of organizing all this information, have been proposed—and it is important to know how far along they have taken us into this alien world, and where they have let us down.

This book is not a typical journalistic review of a scientific realm, introducing the reader to an array of scientists at work in the field or in the laboratory, listening to them chat about their work. Nor is it in any sense an overview of all the kinds of sounds animals emit—a Baedeker to an animal Babel. Instead, it is the presentation of an argument, a line of scientific reasoning that attempts to make plausible, testable sense out of such voices. It is, then, a rather old-fashioned book of science, like those written in earlier times by scientists who were confident that nonscientists could—and would—enjoy following a coherent theory in its entirety.

The word "science" for many people calls up the announcement of discoveries—breakthroughs, as they are often known—typically from the arena of physics and chemistry (a new process that will permit us to make electricity out of cold water, a new ceramic material that permits a rocket nose cone to survive reentry into the atmosphere) or biomedicine (a newfound wrinkle in the human immune system that will permit us to promote health). These are generally considered the hard or, it is implied, precise sciences. In fact, what we typically hear about is applied science. Applied science is a question of how we can manipulate a newfound understanding into something practical. Real *science*—pure science, as it is called by its practitioners with just a bit of invidiousness aimed at their more pragmatic colleagues—is the probing of ignorance with a fairly specific set of intellectual tools. Philosophers of science argue incessantly about the exact nature of these tools and how they are properly used, but the aim of science is relatively clear: to look at the world of phenomena and make generalizations about it that will yield rigorously accurate predictions about how those phenomena will behave next. It is a quest for the true patterns in things and events—in essence, laws of nature.

LOOKING BACKWARD

A presupposition of science is that there *are* laws, ones that connect the simplest of things to the most complex. An evanescent subatomic particle is, we assume, somehow connected to the overall structure and origins of the universe itself. Such particles will behave predictably, or at least within a predictable sphere of probability, and thus the future is not utterly eerie. If this were not the case, we would have no lasers, no H-bombs. Science is a forward-thinking affair. But oddly enough, a good deal of science is a business of predicting backward; we look for the First Event—known as the Big Bang—by sorting through what we conceive to be its aftereffects. We extrapolate backward in geology to see where the continents we live on came from. And the overriding law or principle in biology, which is nothing less than the scientific means for understanding life in all its astounding (and even uncatalogued) diversity, calls for a special kind of backward predicting: evolution.

Every living thing has a past, a set of ancestors. It is far easier to understand why a dog seems to curry favor with its owner or trainer if you understand that the dog descended not that long ago from a group of highly social carnivores, a group of animals in which cooperation was part of survival—its pack. The domestic dog's human family is its pack. In fact, so central is the principle of evolution that a biologist will tell you that if you perceive something about a living creature that makes utterly no sense in evolutionary terms, your perceptions are in error. There are some perceptions that do indeed lead to new refinements in understanding the mechanism that lies behind the process of evolution, but any pattern that is suggested for living things that does not have an evolutionary explanation—that cannot be fitted into the evolutionary framework of life—is almost surely a chimera.

Part of the argument of this book is that many scientific attempts to understand vocal communication among animals have been blind alleys with respect to evolutionary biology. And one reason is that studying animal communication tends to catapult us into our own egos, the

workings of our own psychology, in the sense that we are
so especially good at our own kind of vocal communication that we tend to want to understand nonhuman animal vocalizations in terms of our own talent. We inquire
into animal "talk" because we are the only creatures alive
that do just that: talk. We try to eavesdrop on their conversations with the hidden assumption that those conversations are less elaborate, less elegant versions of our
own. Part of the problem, indeed, lies in our words.
When we perceive a pattern in the utterances of nonhuman animals—even when we see meaning in gestures—
we have few exact words for it. Out of convenience we
call it language, and important distinctions are thus
blurred.

To an evolutionary biologist, it is axiomatic that animal sounds, utterances—voices—are as integral a part of
the animal's being as the shape of its body, the functioning of its internal organs and nervous system, and all the
other behavior it exhibits in the course of its life. All of
this, for each and every animal, has origins without which
none of it can be properly understood. So, having looked
at some, though by no means all, of the vocal communication systems found among animals, and having looked
at some of the ways that scientists have recently tried to
explain these systems, the latter part of this book essentially proposes—and argues—a coherent scheme by
which the origins and thus the meaning of these utterances can be better understood.

Animals communicate in a great variety of ways. Most
of these are nonvocal. A woodpecker can vocalize, but it
also communicates by drumming on trees, a nonvocal

sound. In fact, it's only among land-dwelling vertebrates—amphibians, reptiles, birds and mammals
(whales and dolphins are mammals that left the land for
the sea)—that one finds the use of a voice. And it is this
form of communication—vocal—which is the focus of
this book, though all such animals use other kinds of
signals as well, from special displays to smells—in short,
appeals to virtually all the senses.

Voice—essentially the passage of air through an organ
designed to make a certain range of sounds, a voice
box—is one of the many things we share with these other
(mostly) terrestrial vertebrates. Here we will present several ways by which we can investigate not what is different about each voice but what is common to all. Only
with such an emphasis can we begin to perceive what it
all means. In the process, we hope to shed some light on
how our own unique talent in this arena—speech—may
have arisen from commonly held abilities, and at the
same time provide a new and productive scientific way to
put aside our own anthropocentric dispositions and actually listen. For the key, it turns out, is in the listening.

Most of us get through a more or less satisfactory life
of talking to one another without thinking for an instant
about the nature of the acoustical laws that govern
sound, including the sound of our voices. Even singers
don't need to know physics. But sound is first a physical
phenomenon, and the vocalizations of animals occur in
an environment that itself has acoustical properties, as
does the opera house or the cocktail party. So, inevitably,
the matter of acoustics will come up. So will a certain
amount of theoretical discussion about the nature of

evolution and natural selection, communication and meaning. While most people, we suspect, who read this book will have picked it up out of an abiding interest in animals *per se,* and not in the arcana of acoustics or any theoretical disputations among biologists, we can only say that such matters are necessary to the argument—and to the listening—and we have tried to render them as painless, even as exciting, as possible.

Another caution is in order. The senior author is an ornithologist, which is enough of a reason to expect many examples from the world of birds. But the fact is that, for a variety of reasons, birds have been the most widely studied of animal vocalizers and so it is only fitting that a proper scientific explanation of the voices of nature would lean heavily on these creatures. Nonetheless, the reader will also hear from elephants, whales, mice, bats, squirrels, monkeys and apes—those animals, in fact, that have attracted scientists bent on understanding animal "talk."

As noted, the word "science" typically brings to mind shiny machines of awesome complexity used to probe the very small (quarks and genes) or the very large (supernovae and the Big Bang). Zoology typically brings to mind the fellows who appear on Public Broadcasting (PBS) television specials fooling around with the life in a tidal pool or the hunting behavior of the African wild dog. These are people who study whole animals, organisms, more often than not in the wild. They do not share in the image of The Scientist, and what they accomplish seems perhaps more like the work of advanced naturalists—a kind of soft science, like the social sciences, except maybe even softer because animals cannot answer questionnaires. Someone spotting a character with a parabolic microphone in a swamp may ask, "What are you doing?" On hearing that he is recording the songs of the swamp sparrow, the question will arise: "What is the use of that?" The big money in science appropriations and grants does not flow into organismic zoology: Pouring millions into the analysis of the social life of the chickadee is apt to raise the specter of Senator William Proxmire and his Golden Fleece Awards for silly or petty or foolish expenditures.

And this is a shame. We are all too familiar with the litany of species loss, the destruction of wild habitat, the loss of rain forests and other ecosystems, the unfortunate need to maintain certain of nature's masterpieces in captivity lest they vanish altogether—in short, the inadvertent and advertent sacking of the cathedral of life on the planet. A growing public awareness that something is wrong here is clearly afoot (in part thanks to those PBS specials) and one can sense in the air a longing for a new deal, a new way to think about the value of our fellow creatures. Animal-rights guerrillas, anti-fur-coat referenda, vegans, Earth first! shock troops driving spikes in trees in the national forests to shatter chain saws and milling blades—these are merely the loud and extreme fringes of what seems a widespread upwelling of concern. People are more than ever before seeking an ethical way of contemplating our fellow inhabitants of earth, perhaps the better to arrive at a more supportable ethic for human existence.

In this context, zoology—practiced by those funny

people who wander round out in the boonies looking into the private lives of bugs and beasts—may be one of the most important of the sciences and that is because we have a tendency to dismiss what we do not feel familiar with, or at best to assign the unknown some potential value expressed in human benchmarks, such as dollars. The information supplied through science is generally neutral in an ethical sense. The nature of the atom, the pattern of sunspots, even the evolution of species are all morally neutral facts. It is only we humans who assess values and attach them to this information. We are engaged, it seems, in a grand reevaluation of nature these days—brought on in part out of fear for our own survival but also out of a sense of impending loss of the magic and the genius of nature. We make moral decisions about other creatures every day, ranging from our treatment of our pets to sending contributions to animal preservation organizations, to choosing not to become part of the local zoo's support group, to fertilizing the lawn or spraying for aphids. Many if not most of such decisions are still based on a considerable ignorance of these creatures and where they really fit in our changing sense of values.

In fact, humanity has pursued its own course to the point where we can't say what our place in nature—in the overall scheme of life—is, and many people find this a lonely, even alarming circumstance. Learning to listen to the other voices in nature on their own merits may help to lead us back.

▮▮▮▮▮▮▮

THE MYTH OF DR. DOOLITTLE

For most of the time human beings have inhabited the earth, it has been to them a magical arena, with spirits immanent in every aspect of nature. Communication between these spirits and humans appears to have been relatively commonplace. Shamans (for better) and witches (for worse) frequently communed with their totem animals, took on their spiritual powers, even metamorphosed into them. The idea of talking to the animals hardly seemed beyond human capacity: It was the wisdom and chicanery of animals that made up a great deal of the world view of tribal people for millennia. When the Hopi Indians emerged from underground into this, their fourth world, near the Grand Canyon in what is now Arizona, it was *yaapa,* the mockingbird, that polyglot singer, who gave them their language and doled out other languages to the various peoples of the earth. Ravens, coyotes, eagles, bears—all such creatures have loquaciously informed tribal people all over the world.

For most though not all societies today, this magic has vanished, but it exists at least temporarily for nearly all of us—in childhood. With their fecund imaginations, many children are convinced that they have conversations with household pets along with more humanlike "companions" or with surrogates provided by toy manufacturers. Children's literature has long aided and abetted the dialogue. For several generations, the proficiency of Dr. Doolittle in talking to animals provided sustenance for such beliefs, if sustenance were required. Just when these beliefs vanish is a matter for child psychologists, but that many of us still find the idea attractive is suggested by the stunning popularity (among adults) of the comic strip *Calvin and Hobbes,* the misadventures of an impish, precocious boy and his constant companion, a tiger who reverts to stuffed-animal status only when others are around.

However nostalgic we may be about such fantasies, we generally agree that one is supposed to outgrow them and that grown-ups who continue to believe that their pets talk directly with them are daft. On a recent radio call-in talk show, a woman announced with utter confidence that her dog had spoken to her on a number of occasions in English, and the host used all his diplomacy to get her off the air. There is, however, a widespread belief among people who are not considered daft by most measures that animals, particularly pet animals, "understand" their owners perfectly and communicate specific information nonverbally.

Scientists look on such beliefs as the result of anthropomorphism, the attributing to lower animals of human motivations, goals, desires, and emotions. The more dour among scientists will blame much of this on all that nonsense about talking animals in classic children's literature, television, and the malevolent works of Walt Disney. But, as we will show, many scientists have themselves been trapped in the web of anthropomorphic thinking when they have set out to understand what animals are communicating—if not to us, at least among themselves, in the great cacophony of sounds emitted by the nonhuman creatures in the world.

"Knowing bird language is ever a help," reflected mystery writer Arthur W. Upfield's Inspector Napoleon Bonaparte, when, finding himself in a treetop in the Australian Outback, he overheard a "conference" of crows and could tell from their voices that it concerned something nearby that was dead, and not him. There has probably always been good reason to attempt to "read" the languages of the animals, but none more persuasive than the need to hunt and gather food, or the need to avoid becoming food. The literature of the outdoors—of hunting in particular—is laden with accounts of how the alarm calls of birds and monkeys and other animals have alerted the well-tuned tracker to a desired or undesired presence beyond his sight. In the 1960s, General Electric got a contract from the Limited War Laboratory in Aberdeen, Maryland, to test whether soldiers would be able to tell if enemy troops lay in ambush by listening to animal sounds in the jungle. This is, of course, eavesdropping, since the alarmed animals are not interested in telling humans, be they soldiers or even Mowgli the Jungle Boy, to watch out.

One of the few instances where a wild animal actively sets out to communicate with human beings on a regular basis is to be found among some relatively nondescript birds, the honey guides, for the most part birds of Africa, though two of eleven honey guide species occur in Asia. While mostly insectivorous, they uniquely require wax as part of their diet, in particular beeswax (which a special stomach enzyme breaks down into digestible nutrients). To make access to beeswax simpler, the honey guide has evolved an unusual procedure. It seeks out the assistance of two local honey lovers, the ratel (or African honey badger) and the human being. The bird sets up a great hue and cry of loud chattering until it gets the attention of the ratel or the man. Then it flies off a short distance and chatters again. In this way, it will lead its "assistant" as much as a half-mile to a bees' nest and wait patiently while the nest is torn open and the badger or the man takes the honey. Then it feeds on the wax, thank you very much. (Just how this behavior evolved is anybody's guess. That it is utterly instinctive—written in the bird's genes—and not learned in any conventional sense is attested to by the fact that all honey guides are broodparasites. That is to say, all honey guide eggs are laid in the nests of other species and raised by non-honey-guide parents.)

With a few exceptions, wild animals communicate with humanity only accidentally. Their intentions lie elsewhere. Of course, many wild animals can and do live comfortably in the presence of mankind; one only need think of the pigeons and house sparrows in an urban park. (Curiously, creatures like house sparrows that have become commensals of mankind are very difficult to keep in captivity. Apparently one of the ways they can coexist with us is by becoming very flighty, freaking out when close to a human. They never seem to habituate themselves to close human presence in captivity, hitting the cage walls when approached: quite the opposite of many "wilder" creatures.) Many wild animals seem capable of reading human intentions fairly accurately. Crows legendarily can tell a hunter from an innocent birdwatcher. Lewis Binford, an archeologist who has spent much time living with the Inuit above the Arctic Circle, tells of wolves who evidently know if a man carrying a rifle has any intention of shooting at them.

In any event, with few exceptions, the vocalizations of animals—the songs and calls of birds, the grunts and roars and purrs of mammals, the croaks of toads and frogs, the cricket's chirp, the astonishing array of underwater clicks and booms emitted by fish—all these sounds evolved for purposes that had nothing to do with mankind. They bespeak other business, and it is only relatively recently that scientifically minded people have begun to pry effectively into what that business is all about.

ANIMAL TALK

One

||||||||||

LINES OF INQUIRY

*In which it is suggested that the tools
of science come in many forms, some not
obvious; a useful distinction between
concept and theory is offered; the
elements of language are introduced, and
the authors play a trick on the reader,
whose patience will be sorely tried, but
for ultimately good purpose*

One of the first scientists to consider animal communication in what we would think of as a modern manner
became almost totally discouraged. This was Charles
Darwin, who, in his book *The Expression of the Emotions in
Man and Animals* (1872), concluded that "opposite emotions produce opposite signals." An example he used is
the fearful dog slinking low, tail between its legs, and the
opposite—an aggressive, larger-than-life dog with its
hackles raised. Darwin also speculated on the origins of
voice in animals, noted a number of apparent uses to
which animals put their voices, and then went on to say:

"The cause of widely different sounds being uttered under different emotions and sensations is a very obscure subject. Nor does the rule always hold good that there is any marked difference. For instance with the dog, the bark of anger and that of joy do not differ much, though they can be distinguished. It is not probable that any precise explanation of the cause or source of each particular sound, under different states of the mind, will ever be given."

This is one of the few places where Darwin was so totally off the mark, but of course there was no way he could have foreseen the development of electronic equipment that would increase the capacity of human hearing as remarkably as the telescope and microscope have extended our visual capacity. When Galileo peered through a primitive telescope and discovered the moons of Jupiter, it was a new day in our understanding of our place in the universe. Similarly, when Antonie van Leeuwenhoek, a seventeenth-century Dutch lens maker, peered through the first microscope (a single thick lens, in fact) and saw tiny creatures floating around in a drop of his saliva, it was a staggering new day in the understanding of life. People had begun to see things that no one had ever imagined before. If Thomas Edison and the gramophone had come along a bit earlier, Darwin might not have been so pessimistic about the possibilities of sorting out the meanings of animal vocalizations. But in 1892, even before the gramophone disk had issued forth from Edison's laboratory, a man named R. L. Garner began recording on cylinders at the newly established National Zoological Park in Washington, D.C., by way of inquiring into "the speech of monkeys." Though nothing much came of it, one could say that this was the beginning of the modern study of animal vocal communication. Others before had tried. In the mid-nineteenth century, explorer Richard Burton, a gifted linguist, had listed sixty different sounds made by chimpanzees before giving up the project. Two centuries ago, the gentle naturalist Gilbert White, using a pitch pipe, tried to find out if all owls hooted in the key of B flat. But it was not until the mid-twentieth century that electronic equipment had become sufficiently sophisticated to enable scientists to hear whole new worlds—to listen with the ears, as it were, of the animals for whom animal sounds were designed.

If you pause to think about it, it may seem something of a miracle that most creatures do carry on their vocal lives within *our* auditory range. Why should this be? After all, sound attracts attention—one of its advantages as a mode of communication—but it also attracts the attention of enemies, a defect. One would imagine that most creatures would be better off if they could vocalize outside the hearing range of predators, especially that most devastating predator of all species: us. But of course that would be evolution in reverse. We're the Johnny-come-latelies, and we arose from antecedents that already shared in the auditory range of most other creatures.

It is by no means misanthropic to point out that our hearing and, in fact, our other senses as well are not much to crow about. Our vision, of which we are properly proud, is nothing special compared to that of a starling, much less a hawk. Our sense of smell is weak; there is virtually no way we can imagine the richness of even a domestic dog's olfactory universe.

At this point, it is possible to do little but speculate about what it was like to be the animals that evolved into *Homo sapiens*. Some of the best evidence suggests that they were scavengers for the most part, sneaking to the site of a previous kill once the big predators and hyenas had satisfied their hunger and breaking up the remaining bones to get at the marrow. It is not inconceivable that some of our senses were quantitatively better in those days than they are now (evolution is not necessarily what strikes us as progress), but chances are that they were no better or worse. Instead we probably had to use them more thoroughly in behalf of survival. A blind person's sense of touch and hearing can become astoundingly refined without sight to rely upon. Nevertheless, in none of our five senses is a human being the most sensitive creature.

When it comes to hearing, we have nothing on a barn owl, which can locate a mouse across a barn in pitch-black darkness by its slightest rustle. But, more to the point, we miss a lot of action because much of what goes on in nature simply goes by too fast for us to hear it. Even the music critic who reflexively cringes when a diva strays from perfect mathematical pitch cannot hear the song of a bird—its variety and extent—the way another bird does. Not until this lack was overcome by technological advances did scientists and curious naturalists begin to imagine, much less explore, the complexity of animal vocal communication.

▮▎▮▎▮▎▮

TOOLS AND CONCEPTS

While supersensitive microphones and recording devices and loudspeakers played a crucial role, probably
nothing has been so helpful to listening in on the vocalizations of animals as the sound spectrogram, or voice
print, developed in the 1950s, largely for forensic purposes, by Bell Laboratories. A voice print is as distinc-
tive as a fingerprint. The criminology literature is now
full of such esoterica as "An evaluation of the voice-
print technique of speaker identification" (from the
Proceedings of the 1976 Carnahan Conference on
Crime Countermeasures). A detective-sergeant named
Smrkouski delivered himself of a technical paper titled
"Voiceprints: The uttered truth." Heavy breathers
and bomb threateners have been identified and con-
victed via the voice print: That it was indeed President
Nixon speaking on the Watergate tapes was assured by
this means, not that there was ever much doubt. Sound
spectrograms are literally pictures of sounds and are
indispensable in much linguistic research—for example, in looking into the articulation of vowels, consonants, glottal stops, and the variety and importance of
tone in Mandarin Chinese, as well as understanding the
nature of a soprano's vibrato, various speech disorders,
and normal and abnormal crying among infants—a hu-
mane technology indeed.

The sound spectrograph makes a picture by taking
account of two fundamental aspects of sound—frequency and duration. Duration—how long a sound

lasts—is fairly straightforward. Frequency in sound means waves, an analogy for which is the waves that occur on the surface of a pond when a rock is tossed in.

When the rock hits the water, some of the water has to get out of there to make room for the rock. So a little bit of water is pushed out from this central point where the rock hit. But the rock-sized amount of water that is displaced is not really what makes all the waves. If you were to put some red dye in the water near where the rock hit, you would see that, although the impact of the rock has moved the water around it outward a little, mainly it has imparted to the molecules of water nearby an up-and-down motion that makes the local surface of the water become alternately crest and trough, crest and trough. This motion gets passed on to the water next to it, and so on, all the way to the shore. The up-and-down motion is most intense right around where the rock hit— it is there that the waves go up and down faster and with less distance between them. By the time the waves reach the shore, they have apparently slowed down and the distance from crest to crest (or trough to trough) is longer.

It's much the same with sound. If you shriek, you displace some of the air around you, but mostly you impart to that air a wave action. A high shriek, a squeal, will set the air to moving up and down in place very rapidly—a short wavelength. If you emit a low mooing sound, the air moves up and down in place more slowly and the wavelength is longer. A short wavelength goes past you at a higher rate—updownupdownupdown—which is to say, the waves pass more frequently in a given amount of

8

time. A long wavelength goes by more slowly—up and down, up and down. The crests of low wavelengths pass the same spot less frequently in the same time period. And that is basically what is meant by sound frequency: the higher pitched the sound, the higher the frequency of the sound waves. And vice versa.

(Strategy lesson: if you were an elephant needing to communicate by sound waves over long distances through tree- and bush-covered territory, you would want to emit a lot of sound waves of the longest type, the lowest frequency, because a long wavelength, by virtue of its own shape, is likely to veer around a tree trunk, while a short wavelength is likely to pile right into the tree and bounce off, getting lost in a messy bunch of echoes.)

On a sound spectrogram a sound is broken down into frequency and duration. What we may hear as a single tone—*beep*—may consist of a combination of several different pitches (or frequencies) like a chord on a piano—a combination of notes. The sound spectrograph machine acts like a bank of frequency filters, each of which responds to a particular range (i.e., it only lets a certain range of frequencies through). So it breaks down any sound into its frequency components. A stylus jiggles in response to these various signals and makes an appropriate mark on a strip of paper, drawing a contour map of the sound. Frequency is shown vertically. A high-frequency sound is at the top of the map; low at the bottom. Duration is shown horizontally.

It wasn't long before ornithologists latched onto this machine as a new window on one of the most noticeable

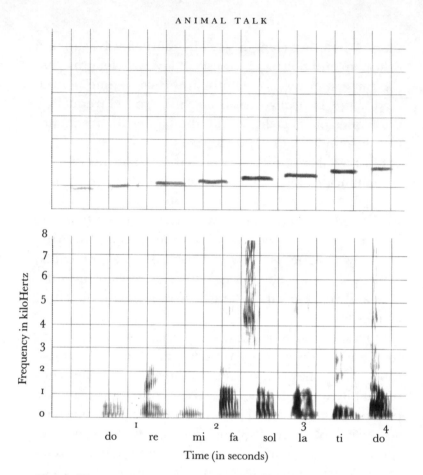

Fig. 1. Two spectrograms made with a Kay Elemetrics Corp. digital Sona-Graph model DSP 5500. Sound frequency, in kilohertz (kHz), runs up the left side, ranging from zero to 8 kHz, or 8,000 cycles (hertz) per second for each spectrogram. Running from left to right along the bottom is time. This scale has four seconds shown. A human singing do re mi fa sol la ti do is depicted on the bottom. Note the high sound produced by the hiss of the "s" in "sol." Above, a human whistles the same notes, which rise as the frequency increases and are thin lines because whistles are tonal, as is most bird song.

things about birds besides their ability to fly—the fact that they vocalize. Today, several popular bird guides, notably the *Golden Guide to North American Birds,* includes sound spectrograms of various bird songs as a means of identifying them, especially when two or more uncooperative species tend to look much alike and otherwise resist identification. It takes a while to look at these contour maps of sound and have them make any sense, but after a while, especially if armed with a helpful tape recording, you get the hang of it. And technology has not stood still here. It is now possible for well-financed ornithologists to buy a computerized "real-time" machine that displays the spectrogram on a screen in color at the same time you hear it. It's like water falling past you.

As important as these new devices have been, the astonishing amount of new information they provide is still mere information, something of a slag heap, until it can be placed into a conceptual framework. And in science there is an important difference between concept and theory, though they are often confused. Knowing the difference is important to the rest of this book, so we might as well get it straightened out now as best we can.

Examples are available from any of the sciences, but we will use biology. In the eighteenth century, naturalist/biologists were seeking to make sense of the kaleidoscope of living forms—both local and exotics brought to their attention in this age of imperialist exploration. A Swede, who came to be known simply as Linnaeus, emerged as the great sense-maker, developing a system for classifying living things that all scientists still use today: those Latinate names that most people can't remember. Linnaeus's system was based chiefly on the degree of similarity in the anatomy (form) and physiology (functioning) of organisms. All plants obviously differed from animals—so, two kingdoms. Within the animal kingdom, there were a lot that had backbones (though more that didn't)—so, vertebrates. Among vertebrates were mammals, and some of these had special equipment for eating plants (herbivores) and others had meat-eating equipment (carnivores). And so forth. Linnaeus was not what we think of as an evolutionary biologist: He lived too early in history. He simply had the *concept* that important similarities in anatomy were the best criteria for putting this big mess of creatures into orderly pigeonholes.

Later, Darwin and others began to see that creatures probably had evolved from one form of life into another over a period of time. This perception of evolution was not really a theory, either, though it is often called one. It is a concept, a general assumption about how things simply are. This perception or concept was greatly facilitated by Linnaeus's system of classification based on similar anatomies. If, on the other hand, Linnaeus had managed to persuade everybody that the proper classification of life forms should be based on size, for example, then there would have been trouble. People would have become mentally accustomed to the idea that trees and elephants had something important in common; the same for herons and tuna fish. Given such a conceptual outlook, it would have been a long time before people would have perceived that life forms *evolve.* And even if they had, they might have invested a lot of time proposing hopeless theories about how little things, like violets, become big things, like giraffes.

A number of people before Darwin, including his grandfather, had conceived of biological evolution. What was lacking was a *theory* of how it worked. And here we get to the difference between theory and concept. A theory tends to be an idea of how something works, and Darwin came up with the notion of natural selection. *That* is his theory. A theory is something you can put to a test. Another theory about how evolution works was suggested by a Frenchman named Lamarck, who said you inherit acquired characteristics, meaning that if you are from a slender family but you work out in a gym and build spectacular muscles, then your children will be born more muscular than you were. Of course, it doesn't work that way. No matter how many Doberman tails the dog breeders dock, Doberman puppies are all born with long tails. End of Lamarck's theory. But the *concept* of evolution still stands.

Concepts are often implicit, unspoken assumptions, but even if unspoken they tend to shape the theories that scientists put to the test. What often happens in science is that when all the imaginable theories fail to work out, we get a new concept, one that suggests new theories to be tested. Geologists used to believe with biblical confidence that the earth was very young—which was then the general concept of the earth—but practically no theory they could devise could successfully explain the facts before them. So it was necessary to conceptualize the earth as a very old place. To understand what goes on in the vocal communication of animals takes theorizing—like any other kind of science— but first it takes an examination of the underlying concepts and, in this odd and complicated field, it is concepts that have clouded our understanding.

▌▌▌▌▌▌▌

THE DESIGN OF LANGUAGE

Given the centrality of language in our own lives, it is little wonder that scientists have tended to use concepts from the study of human language—linguistics—as a means of plumbing the nature of animal communication. One of the most complete systems for making comparisons between the communicative abilities of animals and humans was based on a list of "design features" found in "human verbal language," worked out in the 1960s by linguist Charles F. Hockett. In fact, he meant oral (spoken) human language. Using this list, Cambridge zoologist W. H. Thorpe subsequently rated a number of animal communicative systems, pointing out that all of the design features listed are shared by all human languages and that "each is lacking in one or another animal system of communication."

1. Vocal-auditory channel. From the standpoint of speech, the existence of a vocal tract and of ears seems obvious to the point of banality, but what is not so obvious is that such a channel of communication—as opposed to say, gesture—frees up much of the body for other activities that can be carried out at the same time. Someone conversing in sign language or using a keyboard to write cannot knit a sweater at the same time; nor can a woodpecker fly while it is drumming on a tree.

2. Broadcast transmission and directional reception. Transmission is of course inevitable with sound, but for the receiver to be able to distinguish the direction from which the sound arrives is an evolutionary development not shared by all creatures. A number of fish types, for example, apparently cannot tell the direction from which a sound arrives, their auditory organs being essentially linked too closely together to enable them to do so.

3. Rapid fading. The message does not hang around to be received at the hearer's convenience, like a musky scent deposited along a trail, or a footprint. A message that depends on smell, for example, remains much the same message until it disappears after some relatively lengthy time. The use of sound over the same period permits a variety of messages—even contradictory ones—to succeed one another quickly.

4. Interchangeability. An adult member of any speech community can be either speaker or hearer. Among nonhuman communication systems, for example, courtship displays in spiders, it is often only the males that can send the message while the females do the receiving. In many bird species, it is only the males who sing, though call notes—those little chips and peeps—are usually interchangeable, in the sense meant here.

5. *Complete feedback.* This means that the speaker can hear its own message. On the other hand, in a non-sound communication such as a gesture or a signal conveyed by the colors of one's forehead, the sender may not be able to see the signal, only its effect.

6. *Specialization.* This implies that the use of the system—the vocal tract—to create the sound-bearing message requires sufficiently little energy as to be inconsequential in the overall energy budget of the sender. The only matter of *biological* importance for the sender is the response elicited in the hearer. (As we will see, few singing birds qualify in this.)

7. *Semanticity.* In man, language is meaningful; in this case, meaning is defined as an association between the signal and what is happening in the outside world. (When philosophers are set loose on this one, the whole idea of meaning vanishes. However, for the practical purpose of comparing one system to another or even crossing an urban street, it seems reasonable to think of a signal as having content or meaning.)

8. *Arbitrariness.* The symbol itself (in this context a word such as "elephant") has no physical or geometric identity with the large beast it signifies; the signal is abstract. (As

we will see, this concern has helped lead many interpret-
ers of animal communication astray.)

9. *Discreteness. Astray* means one thing and one thing only:
off course. The slightest change in the signal, like the
addition of an *h,* making the word *ashtray,* brings about
a complete change in meaning. Each word is a discrete
unity. On the other hand, facial displays and gestures by
humans and nonhuman animals usually show continuous
gradation between types: A shy smile can metamorphose
through various gradations into a leer, for example. The
cues by which we determine the content of a smile are
graded, analog (to use computer technology), and subtle. The information in words is digital, or again, discrete. (Even in a word with two quite different
meanings—such as *box*—the context in which the word
is used switches on one or another discrete meaning. No
English-speaking person who hears another say, "Let's
box a little," would imagine that the suggestion means
to make up packages.)

The information in words may be discrete, but the
perception of the *sound* of words, it turns out, evidently
has less to do with what you say in terms of vowels and
consonants and more to do with how the human ear
perceives those sounds. Psycholinguists have struggled
at length trying to figure out just when, acoustically, a
person decides he is hearing *pa* instead of *bah* and it
turns out there is a sudden ("catastrophic") change in
perception rather than any great change in pronunciation. It is our hearing brain that makes the difference. It

is interesting to note, for instance, that in the language of the Hopi Indians the correct English spelling for the words meaning "white person" has never been properly determined. It is either *bahana* or *pahana,* the Hopi pronunciation falling somewhere in between.

10. Displacement. Signals can apply to things or events that are far away in time—either the past or the future—and nowhere nearby or present in space except in our thoughts. Does a robin's alarm call upon identifying a distant speck in the sky as a hawk qualify as something sufficiently far off in time and space to be considered displacement?

11. Openness. A communicative system in which new messages can be coined and understood is considered open and productive. Human language certainly achieves this, as does a chimpanzee that is able to put together a few rudimentary sentences in sign language.

12. Tradition. This means that the conventions of the communication system in part or in whole are passed on by learning and teaching from one group to another. A human child may be genetically prepared for human language in one way or another but, without example, it will not learn to speak clearly, if at all. We are aware of such learning in some birds and primates, and probably among whales, but most vocalizations in nature appear to be programmed into the genes.

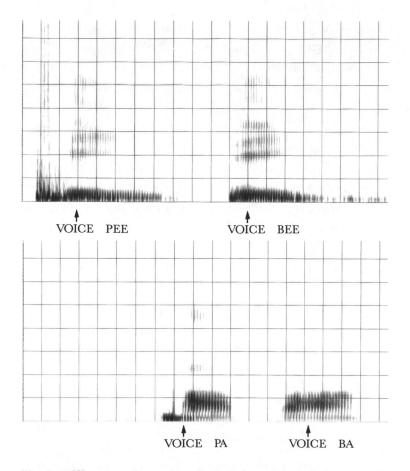

VOICE PEE   VOICE BEE

VOICE PA   VOICE BA

Fig. 2. Different voice onset times of a person saying "pee" then "bee" in the top spectrogram and "pa" followed by "ba" in the lower one. The voice, indicated by the arrows where the vocal chord vibrations begin (the thin vertical lines), is activated much sooner in the "b" sounds than in the "p" sounds. Note also the higher frequencies in the "ee" sounds in the upper spectrogram compared with the "aa" sound in the lower one.

13. Duality of patterning. This means that the elements of a signal may be meaningless in themselves, but in combination, in the form of patterns, they obtain meaning. The individual sounds—vowels and consonants—that are used to make up syllables don't, in themselves, mean anything. And most syllables, in themselves, mean little, if anything at all. Combined into words, all this gibberish takes on pattern and meaning. It is the same with bird song—a collection of individual notes and tones, which, if uttered singly, signify nothing. Put into a characteristic order, the notes bespeak matters that are recognizable by other birds.

These thirteen design features of human spoken language were, for a short time, thought to be sufficient to describe it as an effective communication channel in a social setting, but it soon became apparent that three other elements needed to be added to the list. As it turns out, it is in the application of these three new elements that nonhuman communication systems seem to fail.

14. Prevarication. This is to lie or talk nonsense deliberately, an all too common use of human language and, until recently, not thought to be much used by other species except in certain kinds of play behavior, where "gestures, feints and ruses," as Thorpe said, were "designed to mislead." Since Thorpe's compilation there has come to light a vast number of instances of what has been taken to be animal deceit, particularly among birds and primates.

15. Reflectiveness. This means the ability to communicate about the communication system itself. Thus we have literary critics, communications engineers, linguists, and such maxims as "Watch what I do, not what I say." It is generally thought that reflectiveness is unique to human language. On the other hand, we do know that certain primates and canines can, by their facial expressions or body position, provide a signal that the behavior which follows, although appearing aggressive, is not to be taken seriously. Some people believe that a playful chimpanzee will signal that it means play, not war, by opening its mouth wide and in a relaxed way, the teeth obscured by the lips. Everyone has seen a dog, front legs and chest on the ground, rear end up, tail wagging, launch itself into an otherwise dangerous-looking, growl-filled melee with a playmate. Is that a comment on the communication system?

16. Learnability. This means that the speaker of one language can learn another. Any human being of normal capacity can learn any language spoken by any other human, though the task grows more difficult for almost all people after they have become teenagers. The extraordinary ability of some birds to imitate suggests that humans are not alone in this, but nothing says it so well as the fact that several captive apes have been able to master the rudiments of human sentence structure and meaning to bring about predictable effects. (An irritating question remains, of course; is this *another* language they are learning? Do they already have what could be called a language natural to themselves?)

Thus in a sense, we have a scale made up of sixteen elements or features of human language. Thorpe proceeded to grade ten different communication systems—from those of grasshoppers and signing chimps to human nonverbal communication—for each element, giving a "yes," "no," "partial," or otherwise qualified answer. His chart is reproduced below (Figure 3), and from it we have abstracted (a bit unfairly to Thorpe) an oversimplified bar graph (Figure 4) to show how near and how far animal communication systems are from the paragon (i.e., human speech). The bar graph excludes features one to three and also five, since these are chiefly physical or anatomical matters: The other twelve features might be considered more in the realm of linguistic sophistication/naiveté.

We see from the bar graph that Washoe, a chimp trained to use sign language or some other artificial form of humanlike communication, leaps from a poor seven, based on its own natural vocal communication, to a stunning ten and a half, just a little short of perfection. This chimp/human communication lags only in that it is partly discrete and cannot communicate about the communication system itself. At the same time it is interesting to note how well crows and finches do in linguistic sophistication compared to the relatively crude canine nonvocal system: nine as opposed to four and a half. Dog lovers will be enraged to see that dogs barely outperform crickets, and that doves, with their monotonous cooing and generally dopey demeanor, are the dog's equals.

Of course, the bar chart is all nonsense: arbitrary numerical values assigned to highly qualitative features on a simple-minded basis. Even using a far more complex mathematical system for assigning values would yield the same thing: nonsense. We have adduced this red herring of a graph to emphasize a point about this entire exercise based on Hockett's sixteen linguistic design elements. It demonstrates how easy it is to fall into the trap of trying to understand communication among nonhuman animals from the standpoint of what we know or assume about human communication.

We need to look briefly at one more "system" that was proposed for thinking clearly about these matters.

EVOLUTION AND SIGNALS

Not more than two years after W. H. Thorpe's attempt to make useful comparisons between animal communication systems, William N. Tavolga of the American Museum of Natural History, a student in particular of the variety of noises made by fish and other sea creatures, mused that "the field of animal communication, it seems to me, still lacks the unifying conceptualization for which we seem to be striving." Tavolga complained that the "fact that some species other than man may or may not possess some of these anthropomorphically defined properties of human speech is not helpful as a theoretical framework. Indeed it can be misleading." His point was that a cricket and a chimpanzee are sufficiently distant in evolutionary terms to render any such direct comparisons arbitrary if not absurd.

Design Features (All of which are found in verbal human language)	1 Human Paralin- guistics	2 Crickets, Grass- hoppers	3 Honey Bee Dancing	4 Doves
1. Vocal-auditory channel	Yes (in part)	Auditory but non-vocal	No	Yes
2. Broadcast transmission and directional reception	Yes	Yes	Yes	Yes
3. Rapid fading	Yes	Yes	?	Yes
4. Interchangeability (adults can be both transmitters and receivers)	Largely yes	Partial	Partial	Yes
5. Complete feedback ("speaker" able to perceive everything relevant to his signal production)	Partial	Yes	No?	Yes
6. Specialization (energy unimportant, trigger effect important)	Yes?	Yes?	?	Yes
7. Semanticity (association ties between signals and features in the world)	Yes?	No?	Yes	Yes (in part)
8. Arbitrariness (symbols abstract)	In part	?	No	Yes
9. Discreteness (repertoire discrete not continuous)	Largely no	Yes	No	Yes
10. Displacement (can refer to things remote in time and space)	In part	—	Yes	No
11. Openness (new messages easily coined)	Yes	No	Yes	No
12. Tradition (conventions passed on by teaching and learning)	Yes	Yes?	No?	No
13. Duality of patterning (signal elements meaningless, pattern combinations meaningful)	No	?	No	No
14. Prevarication (ability to lie or talk nonsense)	Yes	No	No	No
15. Reflectiveness (ability to communicate about the system itself)	No	No	No	No
16. Learnability (speaker of one language learns another)	Yes	No(?)	No(?)	No

Fig. 3. W. H. Thorpe's comparison of the communication systems of animals and humans, based on the design features of Hockett.

5 Buntings, Finches, Thrushes, Crows, Etc.	6 Myna	7 Colony Nesting Sea Birds	8 Primates (vocal)	9 Canidae Nonvocal Communica- tion	10 Primates— Chimps, e.g. Washoe
Yes	Yes	Yes	Yes	No	No
Yes	Yes	Yes	Yes	Partly Yes	Partly Yes
Yes	Yes	Yes	Yes	No	No
Partial (Yes if same sex)	Yes	Partial	Yes	Yes	Yes
Yes	Yes	Yes	Yes	No	Yes
Yes	Yes	Yes	Yes	Yes	Yes
Yes	Yes	Yes	Yes	Yes	Yes
Yes	Yes	Yes	Yes	No	Yes
Yes	Yes	Yes	Partial	Partial	Partial
Time No Space Yes	Time No Space Yes	No	Yes	No	Yes
Yes	Yes	No?	Partial	No?	Yes?
Yes	Yes	In part?	No?	?	Yes
Yes	Yes	No?	Yes	Yes	Yes
No	No(?)	No	No	Yes	Yes
No	No	No	No	No	No
Yes (in part)	Yes	No	No?	No	Yes

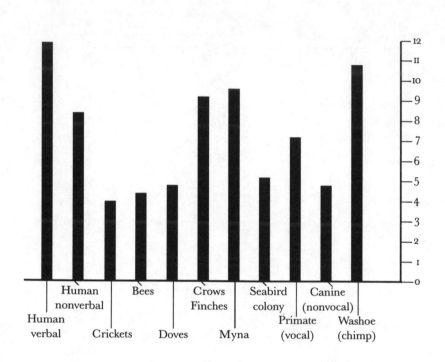

Fig. 4. A bar graph to show how near and how far animal communication systems are from the paragon of human speech shown at the far left.

Tavolga proposed "a theoretical framework for the study of animal communication" (Figure 5) based on the concept of "levels," the levels themselves being based on the nature of the energy emitted (and received). By energy, Tavolga evidently meant a kind of orderliness orchestrated by the animal for the function of getting across what might loosely be called information.

For example, the lowest level of biological energy emission is seen in the mere growth of a plant, each species of plant growing according to its own pattern. An interaction among plants, such as wind pollination, is nothing more than an expression of cellular growth.

The next level up is seen in the jellyfish, an animal with a network of connecting (though not centralized) neurons. A jellyfish is "an integrated whole" and can be thought of as a collection of fairly simple processes, including locomotion, that affects the animal itself and others. Tavolga called this level *tonic.* The "communication" at this tonic level can be specific—that is, a given species gives off a species-specific form of energy. Virtually all animals operate at the tonic level—the scent of a deer is tonic, for example. But jellyfish and sea anemones and sponges never take matters *beyond* the tonic; their only communication with the world is merely the result of their metabolism—of simply being.

On a more complex level, the energy emitted by organisms is directly linked to their physiological and behavioral state. This is called the *phasic* level. Most insects are phasic, as are mollusks, worms, and primitive fish. Phasic outputs are limited in time, and more precise in information content: They take place in connection with certain

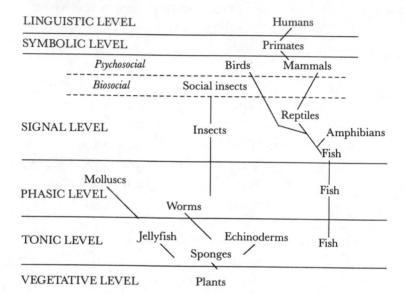

Fig. 5. The levels of organization and interaction of some major groups of organisms, based on Tavolga (1974). Tavolga wrote that the categories (levels) are themselves discrete and "represent quantum differences in behavioral organization and capacities."

functions, like foraging or reproduction. The responses to phasic stimuli are usually "forced," as when a male moth automatically veers off to follow the chemical trail of a female.

Next is the *signal* level. Here, communication takes place by means of specialized devices that produce their output along a single channel. The cold, blinking light of the firefly and the stridulations of the cricket are signals in this sense—both using specialized organs to transmit a species-specific message about their present state of being. So are the croaks of frogs and the call notes of birds (such as short *chips*), as well as more elaborate bird song. At the signal level there are not only highly specialized emission devices—like the bird's voice box—but highly specialized receiving devices called ears.

Beyond signaling, in Tavolga's scheme, lies the *symbolic* goal-directed level, achieved only in higher primates—for example in the "pointing" behavior and other natural gestures of chimpanzees, in human/chimp training situations, and in man. But mankind also and uniquely achieves the *linguistic* level, which Tavolga separates from all other levels by the use of "speech," of "language" derived from "culture" for the communication of ideas. Only mankind spreads ideas around.

At each successive level, from the vegetative through the tonic, phasic, signal, symbolic, and linguistic, the emission is seen to be more specific; more intense, as it were; more differentiated from the general background; and conveying more information. The concept of levels as presented by Tavolga serves to remind us that we are talking about talents that had to have arisen through the process of evolution, as well as coming into being during the actual lifespan of any individual creature. It reminds us that the sponge—little more than a collection of cells—is very good at being a sponge and doesn't need to use elaborate signals to proceed with its life. It need not be invidiously compared to a nightingale. Nor does the nightingale's song need to be parsed by people for whom Mozart is the only proper measure of sonic excellence. Most of life gets along phasically or at what we see as lower levels, and this is worth remembering.

But most of us tend, nevertheless, to be more interested in communications that occur at or above the level that Tavolga labels as *signal.* And it is here that most of the confusion arises in studies of animal communication—confusion that the concept of levels does little to dispel. It is to this vast grab bag level of signals that, as Tavolga says, "one can begin to apply such concepts as design features, communication systems, information content of messages, and semantics." He speaks of "information content" being "coded" at this level, but elsewhere suggests that coding is a property of human speech alone. So we are thrust back into the realm of linguistics and information theory—both of which turn out to be nonevolutionary, nonbiological concepts. These concepts will only take us a certain distance in our attempt to understand the nature of animal communication before either leading us astray or up against a blank wall. For now, it is time to listen to some animals.

Two

SOME VIRTUOSOS

*In which the reader is introduced to
the noisy world of shrimp, fish, and
whales; learns of the voices in his or her
own world that are, to him and her,
inaudible but full of portent nonetheless;
visits a beehive; is provided with what is
(up until this book's appearance)
state-of-the-art knowledge of bird calls
and songs; and looks into the startling
mirror of monkeydom.*

One of the earliest new auditory frontiers to be opened
up was in the oceans, thanks mostly to the exigencies of
submarine warfare. Equipment developed to detect and
track silent submarines began to find that the sea is as
noisy as, in some cases noisier than, our terrestrial world.

A chief cause of this is that water is a better conductor
of sound than air. Sound travels through the air at 1,087
feet per second; in seawater it travels about 5,000 feet in
the same time. In addition, it is reflected (and thus con-
served to an extent) from the sea bottom, from the

boundaries where water of different temperatures meet, and from the surface. Less than one hundredth of one percent of the energy of a sound in the depths gets up through the surface; with such acoustic opacity at this astounding boundary, it is no wonder that naturalists like Jacques Cousteau would dub the oceans "the silent world." But sophisticated underwater listening devices changed all that almost overnight.

Water currents themselves produce a constant background noise comparable to a gentle breeze, a quiet winter night in the country. Added to that are the sounds produced by wave action, the friction of water on the sea floor, the sound of boats and ships, and, superimposed on it all, the outlandish racket of marine animals. All in all, the oceans are noisier than a modern newspaper office with word processors and copiers clicking and buzzing, people talking, phones ringing. It was the need to isolate the quiet sound of a submarine from all this noise that led the U.S. Navy's office of naval research to become one of the major post–World War II funders of research into marine animal behavior and communication. It is a considerable irony that the beginnings of our understanding about the communicatory abilities of those symbols of peace and animal intelligence, whales and dolphins, are directly owed to the rise of nuclear submarines.

THE UNSILENT SEA

What creatures were making all the racket? Crustacea
mostly, and fish, and marine mammals. The loudest of
the crustacea are aptly called pistol shrimp, armed with
a claw about half as big as their bodies, which they hold
in front of them and snap, producing a loud report.
They occur in vast numbers in tropical and subtropical
coastal waters and near coral reefs and, with a whole
population snapping their claws, the noise is comparable to blasts of static on the radio. Evidently, the
shrimp use this high-intensity noise to stun prey, tip-
ping over small fish by the sound and catching them.
The spiny lobster stridulates by rubbing a toothy part
of its antennae over its rostrum (the front of its shell
between the eyes). The shell amplifies the sound, which
can be heard half a football field's length away. A gregarious species of the North Atlantic, lobsters in
groups of up to a thousand will emit constant rasping
sounds, evidently signals that keep the group together
in safety. The same rasping, but faster and thus higher
in pitch, sends them scuttling for cover from an approaching shark or other predator. Females rasp to call
males to them for mating. Several hundred of the 20,-
000 fish species are sound producers. Some stridulate
by grinding their teeth, fins, or bones; others produce a
variety of grunts, squeaks, and drumming sounds by
muscular manipulation of the swim bladder, which acts
as a resonating chamber. In some fish the sound, such
as the grunts of haddock, increases in rate to a steady

hum as the animal becomes more aggressive. The grunts of the male haddock may also stimulate the female to spawn. Thanks to having its sonic muscles attached directly to the swim bladder, the toadfish can emit a sound much like a foghorn; the sound is common during the breeding season. A blast of a toadfish's foghorn is much louder at three feet than the sound of a single-engine, propeller-driven aircraft flying fifteen feet away.

With the possible exception of the white shark in *Jaws*, no sea creatures have captured the imagination of humans so much as whales and dolphins. There is only one Great Novel studied in Western literature courses with an animal as a major character, and that of course is the albino sperm whale in *Moby Dick*. If however, as Ahab thought, that whale was the embodiment of evil, the whale's image has since undergone a total rehabilitation, in part because people in this century began to take note of its increasing endangerment. But nothing has endeared cetaceans to humans so much as the ability to listen in, via phonograph records and tapes, to the mysterious arias of the humpback whales or to hear about experiments that seemed to show that dolphins had the brains and the other apparatus to communicate with great sophistication not only among themselves but perhaps even with man.

Millions of people, flocking to such places as Sea World, know perfectly well that dolphins (bottle-nosed dolphins of TV's *Flipper* fame) can communicate, because they see them performing a variety of feats on verbal command from their trainers. When the ability of dogs and a host of other animals to do the same thing is considered, however, this does not seem so remarkable. Furthermore, dolphins are fine mimics. They are quick to imitate a new sound, and in an ultimately failed attempt to teach them English, John Lilly, an early and famous dolphin researcher, got them to count to ten. But parrots and macaws are far more accomplished mimics, as are mynas. An accomplished starling could give a dolphin a run for the money in mimicry. More recent efforts to "communicate" with dolphins have been based on creating a new "language" of sound symbols—certain whistles, for example—that stand for such concepts as "gate" and "through." Thus, action symbols and object symbols can be used interchangeably, much like the experiments in which chimpanzees can be taught to "converse" with humans by sign language and other techniques, expressing what appear to be desires and intentions. Like chimps, and unlike dogs, the dolphins are known to talk back, repeating the command. Dolphins have been found to have extensive memories for long lists of sounds, and this evident learning ability and the fact that the neocortex— the part of the brain that humans use to reason—covers ninety-eight percent of the dolphin's cortex (in humans it is ninety-six percent) has led many to hope that these creatures possess an intelligence in some way comparable to humans. Which, alas, they do not.

There seems to be evidence that most of the dolphin's neocortex is used to process sound. This is not surprising, since, in nature, they are extremely vocal animals. While research into dolphin-human communication can lead to interesting insights into the extent and the limits of dolphins' cognitive and other abilities we associate with intelligence, so that these limits can be compared to those of other creatures, they do not provide us with any direct information or insight into what the natural vocalizations of these animals accomplish in the course of everyday dolphin life.

Underwater microphones and recordings reveal a vast array of dolphin vocalization that falls into two categories: pulsed, clicklike sounds, and unpulsed groans, whistles, and chirps. Somewhere in between are trains of clicks that have been likened to the sound of a rusty hinge, and a pulsed burst like a Bronx cheer. This variety of sounds serves a variety of purposes. One of the chief functions of the clicking is navigation. A problem for dolphins and their relatives, such as orcas (killer whales) and sperm whales—all of which have teeth—is the murkiness of the ocean in most places. Dolphin vision is good (their sense of smell is almost nonexistent), but a blindfolded dolphin can swim through a complex maze without touching any impediment, simply by emitting bursts of clicks, sometimes as many as 700 a second, and reading the returning echo. They can, by this method, locate an object the size of an apple at 100 yards, read the difference between bone and flesh and even learn to distinguish between two kinds of metal. These bursts of clicks can also be emitted at the upper threshold of sound itself—that is, if they were any louder the excess energy involved would simply turn to heat. This awesome sound can actually be aimed within an arc of less than ten degrees, a kind of sound laser, and based on some limited evidence it is suspected that dolphins use sound as a weapon to immobilize or stun their prey. Loud clicking also appears to occur when two dolphins are threatening each other (even in the most peaceful species, sexual competition raises its often aggressive head). On the other hand, the occasional click appears to be a mechanism for keeping a pod of dolphins together. Much of the clicking is relatively high-pitched and therefore travels less far through the water, sufficient for a relatively small group of animals to apprise one another of their whereabouts. The other sounds that bottlenosed dolphins and other highly gregarious relatives emit—the whistles, for example—are unknown in more primitive species such as river dolphins, where groups are small. The whistles, being lower in pitch, travel farther and conceivably evolved for the purpose of keeping members of large groups in touch. Yet they clearly carry more freight than that.

The murkiness of the ocean is a problem for dolphins—a constraint, biologists would say—but it is also a constraint on biologists. It is difficult enough to catalog the astonishing variety of dolphin whistles and squawks, even harder to associate a particular sound with observed behavior. Some whistles are definite alarm calls: A captive dolphin will whistle when introduced to a new enclosure. There are dialects, too. The alarm call of a dolphin causes members of its pod to disperse; it has no effect on members of another pod. In the clear water off Hawaii, researchers have been able to watch the behavior of spinner dolphins and listen in at the same time. During the day, the animals cruise a home bay, clicking quietly if at all, socializing. As the afternoon wears on, they begin to emit pulsed bursts of sound, apparently an indication of their emotional state, and whistles, which are thought to be statements of identity. Gradually the noise level rises, with whistles taking up more of the vocalizing than pulsed bursts, and the dolphins begin leaping and twisting—hence the name: spinners. There is a great deal of swimming in zigzags, approaching the mouth of the bay and then returning. Bouts of noise are followed by moments of quiet, then more noise, always intensifying until the vast pod heads out to sea for a night of hunting, during which they revert to clicking. Researchers believe that the alternate bouts of noise and silence are one way the pod determines when it is unified and ready for the hunt, similar to the way wolves and hyenas create a prehunting chorus (and which may be taken as analogous to football players in the pregame locker room revving themselves up).

Remarkable patterns have been detected in the vocalizations of the baleen whales, notably humpback whales, which have a repertoire of grunts and squeaks and other sounds but also emit what are called songs. These songs were first recorded off Hawaii by the U.S. Navy and only later identified as coming from humpbacks. Meanwhile, in the Atlantic, personnel listening in for the low-frequency sounds made by large ships or submarines were detecting pulsed sounds so low as to be inaudible to human ears. The sounds were extremely powerful and could be picked up from miles away, and the sources were duly tracked, though unidentified. The military dropped a cloak of secrecy over these findings until the source was identified as the shy finback whale, the second largest of the whales after the blue. Early speculation was that the pulses arose from contraction of the animal's half-ton heart, but it was later found that the sounds were a voluntary form of communication.

What caught the fancy of the public—a recording of their songs became a best-selling LP record—and the attention of scientists, notably Roger Payne, now of Rockefeller University, were the songs of the humpback whale. The songs are sung in the humpback's breeding ground (Hawaii and Baja California in the Pacific and off Bermuda in the Atlantic) and occasionally during migration to those places. They are exclusively a male performance, and may go on for hours; one humpback was recorded singing for twenty-two straight hours. The songs can be extremely complex, consisting of units (each corresponding to a note in music) that are repeated in specific sequences, called phrases, which in turn are combined into what are called themes. Played at high speed, the whale's complex song sounds like bird song, some of which are the most complex vocalizations in nonhuman nature. A song may consist of as many as ten themes, always sung in the same sequence—1,2,3,4,5 . . . and so on, though it may begin at any point along the way. While the Pacific and Atlantic songs are always different, all the animals of one ocean sing what appear to be identical songs at any given time. More remarkable, throughout the breeding season, the songs change—a phrase may be shortened, or dropped altogether and replaced with a new one; one theme may be given a great deal of new material, then stabilized, and another part of the song taken up and altered in its turn. No one knows which whales instigate these changes—is there more than one composer?—but every whale is up-to-date. New parts are generally sung faster. The changes in the song will continue until the end of the breeding season, when the singing stops altogether: At the onset of the next year's breeding season, the old song is recalled precisely as it was, and the changes resume. Within eight years, a humpback's song is totally replaced with new themes and phrases.

To repeat such a complex combination of sounds and to be able to recall them in the same sequence implies an auditory memory system at least as good as if not better than a songbird's, although probably based on a different system. The whales evidently make use of a mnemonic device, perhaps a rhyme pattern similar to those employed by the storyteller bards of old, whose tales were memorized and handed down orally. When analyzing the components—pitch, timing, and harmonics—of the sounds humpbacks make in the course of their songs, scientists found that particular sounds are repeated like rhymes at regular intervals in successive themes. In a song of three themes, rhymes may occur twenty-five percent of the time; in a more complex song, the rhyming doubles. The rhymes may also provide a common thread between old and new themes.

Just why the whales carry on these long, complex bouts of singing is not certain. Some suggest that the songs serve as a spacing mechanism, enabling the males to maintain a proper distance from one another. If one plays the song of a humpback whale back to it over a loudspeaker, it causes a singing male to stop singing and swim away. Others have suggested that it is a mating song, that a female can swim among the singing males and make her choice, much as a female grouse does from the circle of displaying males known as a *lek*. After conception, a female humpback will not ovulate again until her calf is weaned two years later; these unreceptive females actively avoid singing males. Pairing and presumably mating occurs when a singing male spots a receptive female and its calf and joins them as an escort. After some time has passed—a matter of hours—another singing male will try to intrude. The interloper stops singing and a fracas, accompanied by a lot of other noises called social calls, usually ensues. The social calls may bring in up to fifteen more males, all of which try to get between the female and her original escort. Eventually all but one male lose interest, swim off to well-spaced positions, and begin singing again.

███████

ULTRAS AND INFRAS

But it was not just the oceans that held surprises. Biologists armed with electronic listening devices could now eavesdrop on sounds in the air that the human ear could not hear and that had therefore gone largely unimagined. Among the first creatures found to emit such sounds were bats. In the 1940s, using an apparatus designed to detect the high-frequency sounds of insects (that is, ultrasound, meaning too high to be heard by human ears), the young Harvard researcher Donald

Griffin was able to show that not only did bats emit ultrasounds but that they used them to navigate around obstacles in the dark and to locate tiny flying prey. It was a stunning discovery, the first known instance of ultrasonic echolocation in nature: animal sonar. The sounds are emitted in pulses as short as a millisecond. A cruising bat might emit some ten pulses per second, but if it finds itself in a complex spot or approaching prey, such as a flying moth, it will increase the pulses to as much as 200 per second, which gives it a great deal more detailed information about its target. Though amazingly short, the pulses typically start at a high frequency (or pitch) and descend to a lower one, a kind of frequency modulation. The bat's auditory system is capable of detecting nearly infinitesimal differences between the times of the returning echoes of the various frequencies, thus perceiving the distance, size, and speed of the object with uncanny accuracy.

Another kind of bat echolocation is based on the Doppler effect, the principle behind a patrolman's radar, and also the principle behind the abrupt change in sound when an approaching train suddenly passes you. In this instance, the bat emits a single frequency pulse, a relatively long one. If the target is moving toward the bat, the returning frequency is higher than the signal the bat sent out; if the target is moving away, the returning frequency is lower. Bats that use this system are called Doppler bats; the others are called FM bats. Some bats can switch from one system to another as the need arises.

As one might expect, the bat's prey does not take all this lying down. A bat screaming in at 200 pulses a second is making a lot of noise, and many moths can hear the approaching creature long before it can pinpoint its prey. So the moth can take evasive action, flying erratic loops or simply folding its wings and dropping. Others have the capacity for emitting their own ultrasounds, closely matching the bat's frequency and thus, in a sense, jamming the bat's sonar. In such an evolutionary arms race (the moths were around before bats, but bats were flying and echolocating only some ten or fifteen million years after the dinosaurs died out), the bats may be ahead: They simply change frequency, pulsing much higher, and the moth's jamming efforts don't work.

Echolocation is, of course, communication with oneself. And evidently each bat species has its own characteristic sound pattern, to which its brain is specially tuned, thus eliminating confusion in a night sky full of bats and ultrasonic bat beeps. Typically a bat can perceive a target from about two yards away, but it can hear another bat of its own species from as far off as twenty-five-yards or more. Scientists have found that certain bats—those that tend to feed on highly plentiful if localized food sources like blooms of nighttime insects over ponds—are attracted by the sounds of their own kind, evidently listening for those who have found food. On the other hand, the sound of their own kind inspires a threatening attack among bats that feed on scarcer prey: a spacing mechanism. A great deal about bats, including what are surely further subtleties in the vocal system, remains to be discovered, partly because bats, like most mammals, are nocturnal, but most biologists tend to be diurnal.

Bats are said to be the most populous mammals on earth, but rodents like rats and mice can hardly be far behind, and many if not most such small mammals operate at least in part in the ultrasonic realm. Shrews may use ultrasound to echolocate; baby mice emit ultrasound until their eyes open—the sound evidently stimulates licking and nursing on the part of the mother; rats emit a high-frequency whistle when they are losing a battle or giving up, and a higher-pitched whistle when aggressively winning. Male rats emit a two-part ultrasonic vocalization when in pursuit of a female—a warble alternating with a straight tone. Stimulated by the odor of the female in estrus, the male's call may encourage the female to get on with the mating process. Hamster females call ultrasonically to alert males that they are ready to mate, hamsters being largely solitary animals and needing to advertise such matters over distance.

We have been talking about ultrasound, an auditory frontier too high-pitched for the human ear to cross. Generally speaking, we can usually hear sounds running through about nine-and-a-half octaves. A piano has eight. If you made a piano with two more octaves in the treble, you would not be able to hear the last eight or so notes; they would be sounding off in the realm of ultrasound (see Fig. 6). Bats, dolphins, a lot of insects, shrews, and many others could hear those notes. You could not.

Similarly, if you added two more octaves to the bass of this awesome piano, you would not be able to hear the last eight or so notes. They would be vibrating in the realm of *infra*sound, lower than the capacity of the

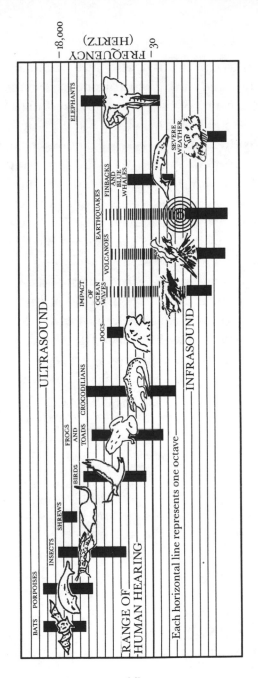

Fig. 6. The range of human hearing, roughly from 30–18,000 Hertz, compared with the sounds of other organisms and sound-generating features of the earth (after National Geographic Society).

human ear to detect. All in all, there would appear to be more infrasound in the world than ultrasound, in spite of the prevalence of bats and mice, and that is because the earth itself is a fairly constant source of infrasound. Severe weather, earthquakes, volcanoes, ocean waves all create infrasound. Indeed even at rest, Mother Earth evidently emits low-frequency groans. Mountain ranges, for example, apparently give off characteristic infrasound signatures that have been tentatively implicated as one of the signals migratory birds use to find their way.

Hindsight suggests that if biologists had found that the smallest of mammals use ultrasound, perhaps they would have guessed that the largest can use infrasound, but it was not hindsight that led to the discovery that elephants indeed communicate a great deal at frequencies below what humans can hear. First there was the mystery, as with bats. Investigators before 1800 knew that bats had an uncanny ability to get around in the dark and one early scientist suspected it had to do with hearing, but the hearing of *what* he had no idea. Modern researchers studying African elephants in the field paid a great deal of attention to the rumbles and trumpeting, the shrieks and barks, snorts and growls elephants use when greeting or signaling to one another, but they also began to notice that elephants or elephant groups that were several miles apart often made synchronous movements, such as simultaneously changing directions or simultaneously pricking up their ears as if listening. Smell could be ruled out (by wind direction) as could vision (by intervening land forms or vegetation). Recently, Katherine Payne, a biologist who had studied the songs of humpback whales for more than a decade, decided to investigate elephants and went to the zoo in Portland to record the sounds of the excellent elephant collection there.

Disappointed at first by the infrequency with which the elephants vocalized, she became aware of a strange sensation; it was as if the entire elephant house would occasionally be subject to an inaudible but palpable trembling, a kind of low-level shuddering of the atmosphere. Returning in due course with the proper equipment, she "listened" in on the previously unknown infrasonic rumbles of the elephants. Later she and William Langbauer took equipment to the field, where they began to unravel the complex infrasonic world of elephant communication.

Adult male elephants live separately from the females and young and spend about two or three months a year in a condition called *musth,* during which they frantically and irritably roam around looking for females in estrus. The females, however, come into estrus only every four or five years and the season lasts only a few days. But as soon as estrus begins, a female is surrounded by males. It is now known that the female "sings" in infrasound; she emits a long, low rumble that rises in pitch and intensity and then falls off to silence, over and over, calling in the males. These songs can be heard for at least two and a half miles and perhaps more. The extremely long wavelengths of such low sounds snake their way through dense woodlands with minimal interference, making it ideal for long-distance communication.

Most of the other vocalizing is done by females, both audible and infrasonic; it is part of the busy family life of mothers, calves, baby-sitters, decisions about where to go to forage, how far to wander, when to head for water. Periodically a female group will "freeze," evidently to listen to signals from a distant group. Meanwhile, the less vocal males "read" the sexual state of various female groups merely by listening in from a distance. Evidence is building up that the elaborate matriarchal society of elephants relies heavily on infrasonic communication to maintain itself over space and time, much as we in our society, with family members scattered here and there, use the telephone.

NONVOCALIZERS

A great deal of animal sound is produced nonvocally. It is only the least populous animals that vocalize—amphibians, reptiles, birds, and mammals. All other noise-makers do it by stridulation, the motion of wings, or other means besides the passage of air from the inside to the outside via a voice box of one sort or another. And most animal communication is not by sound but by means of movement and gesture, color change, odor change, display of feathers or fanny, posture, facial expression, stereotyped wiggles and dances, electric fields, or merely the sound or sight of their passing—nonvocal means. Vocal communication predates the dinosaurs but is a relative latecomer in the evolution of communication among animals. So, although the focus of this book is on vocal communication, it seems useful to point out that one of the most sophisticated forms of animal communication known takes place inside the beehive and is non-vocal. It was discovered by the most helpful tools mankind has at its disposal, close observation and patience, not to mention what might be called genius.

People had already been keeping bees and looking very closely at these highly beneficial economic boons for generations when in the 1930s a German biologist, Karl von Frisch, began to perceive that a successful forager bee, upon returning to the hive, went through what might be called an elaborate ceremony, which, he realized, alerted fellow foragers to the whereabouts of a good food source. This was accomplished specifically by the forager bee waggle-dancing in a figure eight on a vertical part of the hive. Von Frisch discovered that the dancing bee's angle away from the vertical indicates the direction—horizontally over the ground—relative to the angle of the sun. The angle of the dance changes over time to compensate for the movement of the sun across the sky. The intensity of the dance indicated the distance of the food source from the hive. Von Frisch received the Nobel Prize for this work, though a lot of people claimed that his experiments were flawed, that he had not instituted proper scientific controls, that in fact you could get bees to go in predictable directions and distances by simply arranging the odors of the goodies put out for them, and so on. But such objections turned out to be groundless, and by 1989 some of von Frisch's colleagues had created an artificial honeybee, a mechanical imposter operated by computer, and via this apian go-go dancer the scientists actually entered into communication with the bees.

It wasn't the first time they had tried, but they had always failed—often their robots were attacked—because they had overlooked the role of sound in the dance. Typically, it is dark in the hive, so it was assumed that the reactions of other bees to the dance was mostly tactile: They simply touched the dancer in order to receive the information. It later became clear that the dancer exuded not just the odor of what it had foraged on, but often actually regurgitated some of this food. So smell seemed to be implicated, too, perhaps a gauge of the excellence or richness of the foraging area. But soundless robot bees, placed in the hive and dancing with fine intensity, spitting up honey on command, had been either roundly attacked or ignored. It became apparent by 1989 that sound played a role as well—the buzzing of the wings that produces the familiar hum of a bee and usually says, to us, "time for a little avoidance behavior." The dancing bees were also using sound, though nonvocal sound, as part of the system by which they let their fellows know the location, distance, and value of a feeding site. The dancing mechanical bee, still a kind of Van Johnson among Fred Astaires, promises to enable scientists to ask ever more subtle questions about this elaborate system of communication, to tease out the various elements that are involved in what appears to be the symbolic language—even the local dialects—of the bees.

Meanwhile, Thomas D. Seeley, a student of bee communication at Cornell University, has said of the news from the bee robot: "What's remarkable is the ability of these bees to encode information and then decode it through these dances. The dances literally encode information about the distance and direction of a target that can be miles away from the nest." And that is the equivalent of a human providing detailed directions to a spot hundreds of miles away.

The dance of the bees has had profound effects on the way biologists, linguists, philosophers, and just plain citizens think about the nature and sophistication of nonhuman animals. It seems to show that animals as unlike us as insects operate by means of a system that can in essence abstract and communicate data. And this seems the essence of language as we humans practice it. Are the bees, in fact, using language as sophisticated as speech? Are they intelligent in a way akin to what we take to mean intelligence? An 800,000-neuron bee brain hardly has the capacity for what we might consider intelligence (a term of almost spectacular vagueness even in human discourse and comparison): We have some 3 billion neurons in our brains. An insect has far too few cylinders. But a hive full of such creatures . . . perhaps there is some sociobiological, genetic way that a *hive* of bees, powered by the ancient urges of its queen, has something like intelligence . . . ?

This new awareness of the stunning capacity of bees served to redraw the boundaries between human and nonhuman in many minds, as did the virtuoso communicatory performances of birds and our fellow primates, once we had learned to hear them better.

▌▌▌▌▌▌▌

THE LONG AND SHORT OF BIRDS

Birds are generally credited with having the most elaborate vocalizations of all but mankind. This may have developed because birds fly. Virtually every bit of a bird's anatomy—its form—is dictated by the functional requirements of being a flying machine: feathers, light weight, hollow bones, a large-keeled breastbone for the attachment of flight muscles, and so forth. Flight dictates that the size and shape of birds be less various than that of mammals. The size differential between a hummingbird and a swan is far less than the differential between a shrew and a blue whale. There is no contrast among birds to rival the vast anatomical difference between a giraffe and a dachshund. Flight is a kind of straitjacket, and it is also extremely expensive in terms of energy expenditure. As a result, birds tend to fly as little as possible, and those that fly a great deal of the time, like vultures and albatrosses, mostly soar on thermals or wind currents, which is relatively effortless. To extract the maximum energy from food, birds' metabolisms run hotter. The smaller the bird, the tighter its energy budget. Hummingbirds require as much as twice their weight in energy-rich nectar to survive each day, and if they could not damp their metabolisms down at night into a kind of mild torpor, they would not survive till the next day. In such a hazardous energy situation, it is obviously best to avoid occurrences that are energy-intensive, such as squabbles with neighbors, competitors, or predators. A system that could do this, especially over

▐▌

long distances and in forested areas, where visual contact is unlikely, would be of tremendous advantage. A sound that conveys from a long way off the fact that you are here in the middle of your territory and are prepared to defend it can save you a lot of time and energy you would otherwise have to expend flying around the perimeter looking for potential interlopers and threatening them. It is no surprise that the most elaborate vocalizations have arisen among small birds, such as hummingbirds and perching birds (commonly called song birds). Though the perching birds include such birds as ravens, most tend to be small and on tight energy budgets—wrens, chickadees, swallows, warblers, sparrows, finches. In any event, by becoming fliers, birds put a premium on the value of sound and hearing.

Strategic digression: We said that "by becoming fliers, birds put a premium on sound and hearing." This sounds as if there was some deliberate goal on the part of protobirds, near-birds, semifeathered reptiles, what have you, to become the birds we see today, and that they deliberately gave up four-leggedness for a pair of wings and back legs that could hop, and that they somehow actively voted to become more vocal.

Evolution, however, does not proceed in that manner. There have been scientists—the early pro-Darwinist Ernst Haeckel of Germany comes to mind—who postulated a kind of platonic template, say of birdness, or mammalness, or humanness, a grand scheme or endpoint toward which evolution patiently toiled. But both nature's diversity and its commonalities can more easily and readily be explained otherwise. Once a major structural theme, such as a backbone, has come into being in a creature, the descendants of that creature, however diverse, do not give up on the theme. There would seem to be no way conceivable that a jellyfish-like creature could ever arise from a shark (which in fact has a very primitive backbone): Too many options have been passed. On the other hand, over a long period of time some four-legged reptiles became two-legged and winged birds with feathers. The feathers had to have begun as scales that for one reason or another became slightly longer, possibly on the trailing edge of the arms of an arboreal reptile that made a living running around in trees and leaping from one branch to another. These slightly longer scales might have conferred an immediate advantage by making longer and more controlled leaps possible. They might have conferred a primitive form of lift, with each incremental step adding to the creature's maneuverability. Along the serendipitous way to bird-ness, however, other advantages might have accrued; for example, as scales merged over time and countless generations into feathers, they would have had a thermal effect, protecting the animal against sudden changes in temperature. Along the way, a longer leap may have become a glide which in turn became flight, each step producing an increase in the number of niches that creatures with backbones could fill. If there is a certain role that evolution plays, it can be thought of as filling every possible space or ecological situation with life forms. Whether that is also the *purpose* of evolution, part of a deliberate strategy, is a theological question. Meanwhile, the kaleidoscopic variety of life is explainable in terms of incremental changes in the nature of creatures that are weeded out by actual experience, the test being whether an animal manages to reproduce and create young which in turn live to reproduce, passing on the variations. Any new variation that comes about must either confer a present advantage or at least not confer a critical disadvantage.

With very few exceptions, one of them being adult turkey vultures that have no syrinx, or voice box, all birds can communicate by means of calls, often monosyllabic chirps or cheeps or buzzes, brief sounds that are relatively simple acoustically. The young of virtually all bird species have begging calls that help stimulate a parent to feed them. On island rookeries, where colonies of as many as a million seabirds nest in close quarters, a parent can pick out its chick solely by the sound of its begging call. In a similar way, a chick—even an embryo in the egg—learns to distinguish the sound of its parents' calls from those of any other. The infantile begging call may be replaced by another begging call when the bird is a fledgling. The female ptarmigan, an arctic bird, emits a call that has the effect of immediately sending her brood scattering for cover; bobwhite quail mothers have a call that does exactly the opposite—reassembling her young immediately into a line behind her. There are social (or what might better be called spacing) calls by which a mated pair will communicate while foraging, allowing them to stay close enough together but also far enough apart so as not to get in each other's way. Many migrants, especially night migrants, have a particular call, used while flying, that has a similar effect. Aggressive calls are typically low buzzes or growls. A bird may have several different courtship calls (as opposed to songs), and when birds like blue jays mob an unwanted interloper such as a hawk or crow to drive it away, the call to action is a scratchy, low-pitched sound.

A chickadee, perched well up in a tree, will emit an alarm call that is a relatively low *chickadee* if it sees a cat walking along on the ground. The bird is fairly safe, so it doesn't make much difference if the cat can figure out where it is. So the low *dee* alerts nearby chickadees: "Attention!" it is saying. But on seeing a hawk in the sky, a more dangerous situation, the chickadee emits a high-pitched *seee.* The hawk, like all of us, distinguishes the location of a sound by reading the difference in time that the sound strikes one ear and then the other. But if a sound is traveling on wavelengths that are less than twice the distance between the hearer's ears, the location is blurred. There is some evidence that a high-pitched call, such as the *seet* alarm call of a robin, may do more than merely blur the location of the alarmist; it may disguise it in the manner of a ventriloquist. Hawks and owls were found to turn their heads very accurately toward certain sounds, but they were nearly ninety degrees off the mark when it came to the robin's *seet.*

An alarm call may seem to serve the function of alerting one's neighbors or kin to a threat, but, alternatively and, ornithologists think, more likely, it lets the threatening animal know that it has been spotted and may as well give up. (If alone, most birds will emit the appropriate alarm call anyway, sounding the alarm to no one.) Other species often "eavesdrop": The chickadee's *seee* can alert other types of birds in the neighborhood; indeed the alarm calls of many small birds sound very much alike. The calls of blue jays energize virtually everyone around, including deer and squirrels. Certain ducks rely on sea-gulls to provide an early warning system. Purple martins rely on red-winged blackbirds.

It is generally taken as a fact that the various calls of birds—and many passerines have up to fifteen distinct calls, the significance of which can change depending on circumstances or "context"—are automatic responses to the bird's state of mind or surroundings. That is, they are programmed into the bird genetically, are not learned, and cannot change. There could be exceptions; there is some indication that the male goldfinch adjusts its flight call to match that of its mate, and will readjust it to that of a new mate if the first one dies.

Like many animals, such as frogs and puppies, birds will emit a high-pitched shriek when in the clutches of what appears to be a death-delivering predator. In this case, it seems evident that the bird is not calling to any-one. Such behavior seems pointless, but in the long evo-lution of birds, it must be that those that shrieked when faced with death survived more often than those that remained silent. Possibly the shriek occasionally con-fuses the predator, or calls in yet another bigger preda-tor, or one's own kind in an angry mob. Practiced birdwatchers imitate the shriek to bring birds out of the bushes and into view.

Bird calls can serve a variety of communicatory pur-poses and do so without a great deal of energy drain on the bird. On the other hand, bird song—a far more elaborate and often melodious effort—may take a great deal of energy. Percy Bysshe Shelley was right to single out the skylark for special praise—that rarity among birds that "singing still dost soar, and soaring ever singest." Not only is singing a drain on the bird but it exposes it to the attention of predators for what can be extended periods of time. So it must be terribly important. Add to this, as well, the astounding virtuosity of some birds—the brown thrasher has as many as 1,200 recognizably different songs—and you would seem to be confronted with one of the most powerfully expressed needs in nature. Yet bird song, for all its astonishing elaboration, serves what ornithologists believe to be two purposes at best, unlike the several served by the less melodious bird calls: attracting a mate, guarding a territory . . . or both.

Darwin believed that the function of bird song—mostly a male performance—is to attract a mate, the individual quality of the song providing females a means by which to judge the best available male. It has since been found, for example, that the more elaborate a male canary's song, the more quickly a female canary becomes physiologically ready to mate. Logic would suggest that a male bird, having attracted a mate, would then stop singing; and some, like the thrasher, do. But early in this century it began to be clear that many birds—songbirds as well—are territorial animals. Bird song could be seen as a kind of threat/advertisement: "Here I am, and I have chosen this as my territory. Stay away." Logic suggested that the better the territory, the more vibrantly the territory-holder would sing (when there is sufficient food, there is more time to sing), and a female, noting this, would assume that the male has a good territory and join him, actually selecting the territory and not the bird. Taking the process a step further, the male with a good territory presumably "owns" it because of some superior abilities or characteristics, so the female chooses the male indirectly, based on the excellence of his real estate advertising, which is presumed to be truthful. This line of reasoning fits in well with the observed fact that many birds keep on singing well into the reproductive process. At this point, the male has an investment not only in his territory but in his family as well. Savanna sparrows and rose-breasted grosbeaks fall into this category. But then there are birds like reed warblers, which have one song for purposes of territorial defense and another for attracting females. All this led one ornithologist to suggest that there is a continuum of need, with mate attraction (the most important thing for certain birds) at one end and territorial defense at the other, and it is along this continuum that the various bird species are to be found. More logic: A "continuous" singer, like a nightingale, which sings for the entire night with barely an interruption, might be singing chiefly for a mate, while a white-crowned sparrow, whose song is highly discontinuous, may be emphasizing a territorial imperative: The breaks between songs permit the singer to listen for the songs of potential intruders.

Experiments have refined our understanding of the territorial nature of some bird song. In 1925, a British ornithologist, John Krebs, played a classic trick on some great tits (chickadee-like birds of Europe). It was already known that if you removed a territory-holding bird from its territory, its place would soon be taken over by another member of the species, typically a neighbor from a lesser nearby territory. Krebs removed a number of territory-holders, but some of them were replaced with loudspeakers playing their songs, while other areas were left vacant. Still others were given loudspeakers playing a similar tune on a tin whistle. Great tits from other territories almost immediately invaded the areas with the tin whistle or no song at all: They remained clear of those where actual tit songs were being played for a much longer time. The song, and the song alone, was sufficient to keep interlopers out.

Such an uncompromising and antisocial role for nature's most melodious sounds hardly seems in accord with our aesthetic reaction to bird song. One would prefer that it was all at least as romantic as Darwin suggested. One can speculate that our earliest ancestors listened to the dawn chorus of the birds and took it as joyous proof that the sun was yet again going to rise and that life would continue for another day. Similarly, while the magical appearance of migrants was surely interpreted as the official onset of spring, many birds that overwinter in northerly climes begin to crank up their songs well before the snows of February have begun to thaw. The sun plays a direct role in such affairs, serving as a great light-bearing metronome that, in combination with the shift of the earth's tilt, stimulates the birds' hormonal shifts, bringing on increases in the sizes of sex organs and the behavior, including song, associated with such growth. We have relearned, through scientific observation and experimentation, what less sophisticated peoples have always known—that birds and other forms of life are part of a global, interdependent web of things. Within this overall insight (expanded from the localized world view of the hunter and gatherer to the planetary view of science), the chief difference is that scientists see cause and effect in such affairs in a rather different light.

Why the dawn chorus? Presumably most birds have spent a relatively long night without refueling, and given their demanding metabolisms, one might expect them to go look for breakfast rather than carry on in self-advertisement first thing in the morning. The dark, or near dark, however, is a poor time for diurnal birds to forage—vision typically being the key sense involved—and at such times birds look for food desultorily at best. Ornithologists have reasoned that since one's neighbors are not fully engaged in looking for food at such times, the temptation to sneak into a superior territory might be even greater then (an avian version of the maxim that the Devil makes work for idle hands). If so, this would be an important time to inform potential interlopers that your territory remains inhabited.

As birdwatchers know, midday to early evening is dead time. It is a time of great stillness, when the birds seem to rest and, but for a hawk or vulture circling overhead, there is little to see or hear. Toward the end of the day the chorus resumes, perhaps because the oncoming dusk is another time for illicit forays. Similarly, there tends to be an annual cycle: After the mating season, most bird song tends to disappear, but with the beginning of autumn many birds begin to sing again—a quieter version of their spring song, accompanied by a slight reswelling of the sexual organs. It has been proposed that this period serves as a kind of practice session for young birds, a time when fledglings who will have to join the fray of territorial battle the following spring learn to get their songs down right.

Just how birds learn their songs, in fact, has been a subject of enormous interest to ornithologists. (And even into the 1980s, long after the humpbacks had been eavesdropped on, it was fashionable for ornithologists to say that, besides mankind, birds were alone in this sort of learning.) For many birds—like the northern oriole or the cardinal among common backyard species—the range of songs is learned by the end of the bird's first year. After that, few if any additional variations are learned. For others, such as the mockingbird, life is a continuing course in creativity, the bird working constant variations of what might be called syllable and phrase on the basic species theme. There is evidence that for many birds, including the mourning dove and all flycatchers, the song is basically wired in genetically, while others need to hear their own kind singing. High-quality instrumentation has allowed scientists to tune in so precisely that it has been possible, for example, to establish that young house finches pick up their species' song by listening to other males of the species in the neighborhood rather than to their fathers. On the other hand, birds such as the European bullfinch have been shown to learn *only* from the father. Bird song can be something of an address: Part of the song may say "species," another "locale," and yet another "individual."

There are a number of ways to look directly in on the

learning process. One is to deafen young birds by severing the auditory nerve. A pigeon that is deafened will develop its species's call sounds perfectly, as will a chicken or turkey. Left with their hearing intact but raised by foster parents of other species, they still develop their own characteristic vocalizations. On the other hand, cardinals and white-throated sparrows, to name only two songbirds, will sing a barely recognizable version of their own song if raised in silence, though this poor version is closer to the norm than if they are actually deafened early on. This suggests that, while in some birds whatever forms the neural template for the species's vocalization is complete at birth, others must refine the template to some degree by listening to themselves, and perfect it by listening in on others.

The more scientists have listened in on bird song, the more complex it seems to become, which is not altogether surprising given the fact that there are some 9,600 species of birds in the world (nearly twice the number of mammal species) and nearly half of the birds are capable of song. Once song has become established in the course of evolution as a means of vocal communication, it is reasonable to expect variation to proliferate just as singing species have, since each species inhabits a particular niche, an ecological realm that is, however subtly, different from that of any other species. And if the raw material of evolution is genetically determined (including the inherent capacity in some species to learn a song), the context is the environment. Bird song, like any vocalization, operates in a particular environment that presumably places restraints upon it. As an advertising tool, song is a good long-distance medium, often permitting a bird to be "known" without being seen. But volume isn't everything. A high-pitched sound, as we have seen, travels in shorter wavelengths that can easily bump into trees and be deflected: Birds such as the ovenbird or wood thrush, which spend a great deal of time on the ground, sing with a low-pitched wavelength, the longer wavelengths being able to go around trees and other obstacles. Yellow warblers, which spend most of their time in the shrubbery, have a higher-pitched song, and the bay-breasted and Blackburnian warblers, which carry on in the tops of trees, sing at the highest pitch known among warbler species.

If you watch an opera diva sing—even on television with the sound off—it is immediately apparent that a great deal of the music is shaped by what she does with her mouth and tongue. The human vocal system is in a sense multistoried, beginning with the larynx, located toward the upper end of the windpipe or trachea. The muscles of the larynx control the tension of vocal cords, which are moved by exhaled breath and produce sounds. These are modified by the resonating capacities of the upper vocal tract and finally shaped in many ways by the apparatus of the mouth. For example, if you take a deep breath and then emit a single high note—*Eeeeeeeee*—your lips will most likely be drawn back in an expression approaching a grin. But if, as you continue to emit the *eeeee* sound, you change the shape of your lips into a tight circle, as if preparing to kiss your aunt on the cheek, you will hear your *eeeeee* gradually turn into *eeeeeeeeyooouuu-uuu*.

Birds cannot do this. The upper part of a bird's vocal tract may play a role as a resonator—hence the characteristic movements of the head and neck of a singing bird—but for the most part the music is produced in an organ unique to birds, called the syrinx. This is typically found at the point where the trachea branches out into two tubes that go to the lungs—the bronchial tubes— and it consists largely of membranes with pairs of muscles. The muscles control the tension of the membranes so that air passing over them creates a variety of vibrations, and the number of syringeal muscle pairs evidently has much to do with the complexity of the sound produced. Storks and ostriches have a syrinx but no syringeal muscles, and so are restricted to hissing and grunting. A pigeon has one set of syringeal muscles; starlings, mockingbirds, and crows—virtuosos all—have between seven and nine pairs. Some birds, like the wood and hermit thrushes, which sing what many consider among the most poignantly beautiful songs of all, can sing two songs at once, controlling the left and right sides of the syrinx separately to create a harmonic melodiousness unmatchable by the human voice.

WE PRIMATES

It is, nevertheless, the human voice that is predominant on Earth and this is surely because of the profound intellectual step made at some point in human evolution: the development of human speech, the properties of which have been the subject of intense study for well over a century by scientifically minded linguists and others. The two most important distinguishing features of human speech (and linguistic capability) seem to have relatively little to do with the actual range of sounds we can emit—vowels, consonants, back-of-the-throat glugs called glottal stops, clicks, and whistles. What seems most crucial is that human beings are predisposed genetically to human grammar—for example, sentences that have a subject and a verb—and that most of the sounds we emit are totally arbitrary. Every human language has a different-sounding word for "man" and "woman." It is these qualities that give humans the ability to comment on matters that are not present—in space or time or both—an ability denied to other animals. Such an ability must have evolved from the capacities present in our more immediate forebears, but since our hominid predecessors are with us only in the form of fragmentary fossils, it has seemed reasonable to look among other living primates to see how close they may have come to developing some of the properties unique to human speech.

As noted earlier, experiments involving sign language and other visual symbols have shown that chimpanzees, and in at least one instance a gorilla, are capable of a rudimentary *use* of human grammar. Like dolphins, they can sometimes put together arbitrary symbols with specific meanings—verbs and nouns—into simple sentences, even evidently referring to events that occurred earlier or objects that are not present. As a British anthropologist recently pointed out, most of these primate studies have been done by people of an anthropological bent, providing them with a not unreasonably biased interest in human affairs: Their subjects, the chimps and gorillas, have tended to be viewed as lesser versions of humans, and often compared to human children. It is only recently that zoologists have gone back into the field to see if what appears to be a rudimentary intellectual capacity for something akin to human grammar is ever put to use in the daily life of chimps in the wild—that is, without training by humans. Can a chimp alert its fellows to the existence of a snake that it saw "over there a little while ago?" So far the results are inconclusive. But there has been a great deal of interest in animal communication circles over the past decade about the precision with which a certain species of monkey announces the presence of different kinds of predators. It is as if they used words.

These are vervet monkeys, medium-sized monkeys that live in fairly large troops in sub-Saharan Africa, inhabiting savanna forest areas. They are subject to predation or danger from a variety of animals, as are most creatures, but it began to be clear that they make precise distinctions in such matters. Leopards, for example, hunt in daytime, lurking in bushes to snatch a passing monkey. Though good climbers, once in a tree leopards aren't quick enough to catch a vervet. When a leopard is spotted by a member of a vervet troop that is foraging on open ground, one of the troop will emit a particular bark and all the others immediately take to the nearest tree. On the other hand, an eagle can pluck a vervet monkey off the ground or out of the tops of trees. Upon spying an eagle, a vervet sounds an alarm more like a chuckle than a bark and all the vervets head for the densest bushes nearby. If a vervet comes across a snake, particularly a python, it sounds a high-pitched chittering alarm and the troop all rise on their hind legs and look at the ground around them. Recording these sounds and then playing them back, Dorothy Cheney and Robert Seyfarth have studied the specificity of these calls, suggesting that they are used in the same way words would be. In fact, the calls are not as precise as human words: The leopard call serves to alert the troop to the presence of lions, hyenas, jackals, and cheetahs as well (all of which call for evasion by climbing trees). Similarly, four different kinds of eagles, two of which are less common threats, are specified by the eagle call, and the snake call refers to a variety of large and dangerous snakes.

Is this inborn or learned behavior? Evidently young vervets are predisposed to giving alarm calls that refer to large *classes* of threats: The leopard call is given by young ones for anything that walks on the ground and makes them uncomfortable, including walking birds like storks; the eagle alarm may even be given at the sight of a falling leaf; the snake alarm can draw attention to a vine as well as a range of nondangerous snakes. Eventually the young learn to narrow these calls down to legitimate threats in the three different classes. At the same time, the young will sound a different alarm on seeing a baboon; baboons will prey on infant vervets, but adult vervets, which are too big for baboons, are not threatened and do not emit a baboon alarm on seeing one. The young vervets evidently learn from their mothers which predators are dangerous and which aren't. If, for example, a young vervet hears the eagle alarm, it will look at its mother. If she is looking up, the young one will also look up. Before long, it learns to look up immediately at the sound of such a call. And if the mother looks up and then flees for the dense bush, the young one soon learns this is the right response. But if a young one gives the eagle call and its mother looks up and does nothing (since the flying object was a harmless pigeon), the young one eventually learns that pigeons don't warrant the call. Not only do young vervets learn the traditional calls, the monkeys themselves apparently can learn new signals for new situations. Rather late in vervet evolution, one imagines, human hunters came along with dogs, and in this situation, a vervet emits a soft alarm call barely distinguishable from the usual background noise of the environment. Its fellow monkeys respond to this soft call by heading for the bushes, where humans and dogs cannot follow. Are the vervets giving us an example of the origin of words? It seems difficult, on the surface, to doubt it.

Three

||:|:|:|:||

MORE LINES OF INQUIRY

In which the engineers are allowed their say (briefly); the hidden anthropocentrism of biologists is relentlessly revealed; the origins of peacock tails and much nonhuman choreography is made logical; the elegance of natural selection is made clear; and the reader is given a welcome moment to pause and reflect on what has so far been explored.

We humans tend to be rather pleased with the superior nature of our language ability—as indeed we should— but as anyone who has been to a college seminar or two knows, our language can fail when it comes to providing us with a way of understanding certain concepts or ideas: communication, for example. In its simplest etymological sense, communication means the act of sharing, even the simple *fact* of sharing—communicating rooms sit there rather passively doing nothing at all but sharing a door. At the other end of the communication spectrum lies the car salesman persuading people to part with more money than they had planned for automobiles they may not need or, in pop psychological terms, a couple (call them Frank and Trulie Earnest) working with all their verbal might to achieve a perfect understanding of each another.

It has been fairly well known for a long time that many plants, such as trees, have evolved with chemical substances in their edible parts that repulse or deter those creatures, particularly insects, that want to eat them. More recently, however, it was found that at least some trees, when attacked by such a pest, will emit a substance that, upon being borne by the wind to other similar trees, stimulates *them* to produce extra quantities of the substance that discourages that particular pest. It is obviously safe to say that the communication was unintentional in that it is hard to imagine trees having what we think of as intentions—momentary or even long-term goals. Of course, there lies in the tree's cells a molecular arrangement (called DNA) that bears a kind of program for the development and survival of this tree, based on a long series of experiments by its ancestors, and including the manufacture of nasty chemicals in its edible parts the better to ward off the sort of pests that its kind has encountered over the generations. But that's different from having an intention, although just what advantage it is to one tree to let others of its kind know about the arrival of a pest in the neighborhood is an interesting question. But the matter at hand is not biological logic so much as the problem we have with pinning down such concepts as communication, even given the richness of our language. For example, if a protozoan abruptly changes course upon encountering an unpleasant increase in the temperature of its environment, is that a case of communication? Certainly it was the receipt of what might loosely be called information that brought about a resulting change of course, of action. Biologists call such automatic responses to a stimulus *taxis*. Some creatures are *phototaxic*—literally "moved by the light." An increase or decrease in the amount of light will bring about a change of course. But so automatic a reaction—similar to the way you close your eyes tightly when a brilliant light flashes—hardly seems to qualify as communication.

What are we to make of a female moth trailing a tiny stream of airborne molecules—a chemical substance, almost like a perfume, called a pheromone—which *in effect* says to a male moth (1) I am of your species, (2) I am a female of your species, (3) I am ready to mate, and (4) if you follow your nose upwind you'll find me. Is that too simple to be called communication? We cannot imagine the sensation in the male moth when it first receives such a message. Certainly it does not rehearse anything like our numbered sentences above. Its reaction is automatic; it has no choice but to veer off and follow the pheromone trail. The urge, though great, could shortly be superseded by a less-attenuated perfume trail, or the sound of a moth-eating bat. The success of the commercial perfume business over millennia should be all the testimony needed to prove that a large number of human beings believe that such chemical signals are an effective form of communication. But it can be argued that there is nothing automatic in using perfume, that the wearer has a fairly specific goal, a formulated purpose, in mind, and sets out to encode his or her intentions (in this case in the form of a scent artificially applied) in order to alter the behavior of someone in particular (or for that matter persons yet unknown). Does the moth exuding phero-mones—an activity that has obvious *utility*—and the lady applying perfume with a *purpose* constitute a difference in kind or merely two points along a continuum of a form or medium of communication?

In any symposium on the topic of animal communica-tion, one is likely to hear as many definitions of commu-nication as there are speaking participants. In one such meeting, a communications engineer created so strin-gent a model of communication involving goal-encoding intended meanings and decoded understandings, affect-ing the perception of facts, choice of skills, and so on, that a biologist later observed that communication had been taken away from all creatures but mankind and perhaps a few chimpanzees trained to use sign language. (He could have added computers properly hooked up and programmed.) Such a definition can help the analy-sis of animal communication by focusing attention on the totality of fine details of whatever system of signals is under scrutiny, but it may make an implicit assumption that communication among nonhuman animals some-how takes place according to human standards and con-ceptions. Even so, one must start somewhere, and most students of animal communication have started with the assumption that animals produce or emit something we can call signals, and that the signals bear something called information.

■I■I■I■

INFORMATION AND LANGUAGE

We tend to think of information as something that exists by itself (and maybe in our human universe it does). We are inclined to think that animals, including ourselves, transmit this "information" and that its communication contributes to the orderliness of interactions and the maintenance of social structures. Information theorists tend to wax mathematical and tell us that information can be quantified, in the sense that a transmission thereof will reduce uncertainty in the receiver by a measurable factor. That is, the recipient of information is seen as essentially sitting around with a lot of options. Say it could go in any of the four cardinal directions. A signal comes in and says either (1) everything east, west, and north is terrible, or (2) south looks good, which is not the same as (1). Both signals could lead to the same action but not necessarily at the same level of uncertainty for the recipient. The first signal—that three directions look bad—has more solid information. By the same token, if one coin was weighted to have a 70/30 chance of coming up heads, and another had a 50/50 chance, the toss of which coin would yield the most information? Answer: The 50/50 coin, because more uncertainty was reduced after it was tossed compared to the toss of the coin with a greater chance of coming up heads (this logic is the origin of the use of binary digits—*o*'s and *1*'s or bits—in computers. A 50/50 probability contains more information than any other).

The male moth basically gets one signal at a time. It tells him, for instance, that south is good—though he knows nothing of the other directions. But over a short period of time he may be getting signals from hundreds of other points in a sphere of directions, perhaps all good and none of them in the slightest bit abstract. They are as tangible as a tap on the knee by the family physician. The male moth is choosing from many such taps over any given length of time.

We humans are given to a great many "knee-jerk" reactions to stimuli, but we receive very little of what we generally conceive of as information this way. Instead it comes through abstract symbols that need to be considered. Try this little experiment:

Say the word *parpar.*

Now say *mariposa.*

Papathi.

Schmetterling.

Papillon. (A book and then a movie, right, about a guy who escaped from Devil's Island?)

Now there is nothing in the *appearance* of those words on the page, or the *sound* of them as you pronounced them, however accurately or inaccurately, to suggest what we (in English) think of as a butterfly, a member of lepidoptera, an insect with big colorful wings that flutters around prettily from flower to flower, and which nobody is scared of. The ancient Egyptians, who used hieroglyphs, and the ancient Chinese, who used pictographs, might have written "butterfly" so that you could actually *see* what they meant. The written word (especially when it ceases to be a picture) and the spoken word are both abstractions, neutral collections of sounds or shapes that have assigned and agreed-upon meanings. Sorting this out is the task of every human child, first the sounds and then their assigned meanings, and children tend to be strikingly competent at this. Learning to read is harder. Indeed, it has been found that human children are genetically predisposed to take the abstract association between sound and meaning for granted.

Of course, some words do have a direct relationship to the meaning they designate. They are audible "pictures." *Ding-dong. Thud. Ring.* This is called onomatopoeia, and poets are especially fond of it. Edgar Allan Poe, dissatisfied with the word/sounds for the ringing of bells, invented his own: tintinnabulation, from the Latin word for "bell"—*tintinnabul(um).*

HIDDEN ASSUMPTIONS

It is the very power of our own talent for speech, and our linguistic conception, that informs another approach to animal communication commonly taken by most students. In this approach, one breaks apart the "components" of communication into three separate levels: A series of tacit assumptions is made about the nature of animal communications—assumptions that have long made it nearly impossible to understand what is going on when animals employ vocal sounds. The first level is called the *syntactic,* where one looks at the signals themselves simply as physical entities, apart from any function they may or may not serve. The tacit assumption here is that the actual sound—be it the hoot of an owl, the roar of a tiger, the grunt of a chimpanzee—is of no relevance to the meaning or lack of meaning in the signal. On this level, the signals are viewed like the 1's and 0's on a computer printout. A great deal of information can be conveyed in a large enough sequence of 1's and 0's, but the choice of 1's and 0's to convey information is largely arbitrary; it could just as easily have been 1's and X's. One question that can be asked at a syntactic level is: Is that particular sound signal capable of conveying a lot of information or only a little? For example, a monotonic click that can be emitted only at one level of loudness cannot be imagined to be able to carry a complex message, any more than someone could produce a concerto by cracking his knuckles. On the other hand, a sound signal that in fact consists of many notes at many frequencies and many levels of loudness can, theoretically, carry a good deal of information.

And information brings us to the next level, the *semantic.* It is at this level that one begins to look into the meaning of the signal, the information that is by some convention or conventions within a species somehow encoded into the otherwise arbitrary sounds its members emit. Beyond the semantic level lies the *pragmatic* level, which is to say the effects the signal has on the hearer. The assumption here is that the information that is encoded in the signal can be decoded by a hearer and acted upon.

Some biologists have tended to interpret "information" as its dictionary counterpart: knowledge. And since the development of information theory for the use of computers, others have interpreted in the more mathematical, rigorous sense of reducing uncertainty on the part of the receiver of the information. One of the results of this from the standpoint of definitions is that if you have an animal paying no measurable attention to another animal that is frantically shrieking *seet!,* the information encoded in the call did not turn into what we call communication. Faced with the fact that the absence of response to vocalizations is exceedingly common, the biologist has to assume that information can exist and be transmitted without being received. Another result—far more important from the standpoint of understanding vocal communication among animals—is that the assumption that the *information* in these calls is some kind of independent entity that causes things to happen led scientists to another unspoken assumption: that animal communication could be studied by itself—without any particular regard to the rest of what goes on in an animal's biological existence.

For example, suppose we take the tail-wagging behavior of a squirrel confronting a snake and try to find something quantitative about it. We could quickly come up with a series of testable hypotheses having to do with the proximity of the snake, the age of the squirrel, the density of squirrels in the neighborhood, the breeding condition of the squirrel, the size of the snake, the size of the

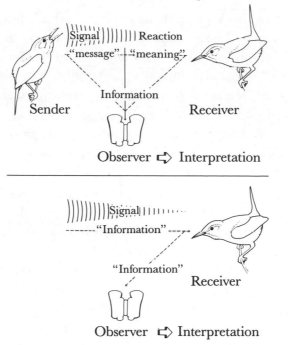

Fig. 7. The two ways that observers derive information inter-
pretations from animals communicating. Above, the linguistic
approach focuses on both the sender and the receiver. The
sender's "message" is adduced from describing all the situations or contexts in which a particular sound is given. For
example, if a sound is given most often when a bird is about
to land, or when slowing or veering in flight, its message would
be called *locomotory behavior*. Messages are concerned with the
kinds of information shared by senders. Meaning is adduced
by what responses the recipient makes, what the information
meant to them.

    Below, information theory ignores the sender and concen-
trates on the changes in receiver behavior, its "reduction of
uncertainty," upon receipt of a signal. In both linguistic and
information theory approaches, the information in the signal
is assumed to have caused the receiver to behave as it did.

tail, and so on. We might discover that the intensity of tail-wagging—the rate and extent of tail-wags—has a direct correlation with the proximity of the snake to the squirrel. Because we subscribe to the informational perspective in such matters of signaling, we would be satisfied that we had decoded the message of tail-wagging in squirrels. But perhaps there is more to it than that—or less. Tail-wagging in squirrels could have to do with something else besides (or as well as) snakes. The tail-wagging could be directed at other predators that might surprise a squirrel that is concentrating on a snake—for example, inducing an attacking hawk to go for the tail rather than a more vital part of the body.

Yet another effect of this informational assumption was that it now became possible to assume that each animal species has its own distinct language, made up of arbitrary sounds. Another tacit assumption was that the hearer of a vocal communication is the passive partner in the proceedings—receptive or not as the case may be, but not an active part of the equation.

As it turns out, none of the tacit assumptions in this linguistic model of vocal communication in animals is faithful to the real world in which animals operate. Until recently, hamstrung by the anthropomorphic concepts of linguistics and information theory, biology had not produced its own framework for studying vocal communication in animals. The implicit notion that each species has its own unique language long precluded the discovery of any generalities or patterns in vocal signals across species lines. The animals that produce vocal sounds, as opposed to stridulations or clicks, are amphibians, reptiles, birds, and mammals, all of which share a common ancestor—a fishlike creature that took to living part of its lifespan on land. The gills of that fishlike creature evolved into ears. It would be surprising if there were not some commonalities, some generalities, among the signals of all vocalizers, but the informational perspective led scientists dealing with vocal behavior away from the great organizing principle of life—evolution.

PEACOCKS' TAILS AND LOGIC

A special branch of zoology is called ethology, the study of behavior characteristic of animals. A tenet of ethology is that behavior, like the organs of the body, evolves. On the other hand, behavior rarely shows up in the fossil record, so to trace the evolution of, for example, a courtship display takes another kind of comparative study. The sheer visual magnificence of the peacock's fan (it is not actually his tail, but an elaboration of feathers called tail coverlets) can overshadow in our minds its actual use in the ritualized courtship display of these birds. When a female appears, the male raises the fan (and his tail) and turns away from the female, rattling his tail feathers against the fan and taking short mincing steps backward. When the female comes into view around the side of his fan, he arches it farther over his head and adopts a rigid pose with his body tilted forward and his head bent. If the female crouches before him, mating takes place. Where did this stylized behavior come from? It came from ancestors, and although there are no peafowl ancestors around now, there are other species of pheasants that are distantly related. In many pheasants, too, the courtship display includes taking up a rigid pose with the body tilted forward. In some, the male pecks the ground while in the tilted position, and the female may run up to see what he is pecking at. At the other end of the series is the domestic rooster, which calls the female to feed, as do many other birds, before mating. Scientists can reasonably opine that there was also an evolutionary continuum from visible feeding that was progressively overlaid with stylized behavior until, in the peacock, the feeding behavior disappeared. The behavior has become stereotyped and exaggerated—characteristics that serve to make it highly conspicuous against the background noise of motion and normal daily behavior. Ethologists have spent some time working out the evolutionary origins of visual displays from such nondisplay precursors as feeding behavior, preflight behavior, and the like, but the origins of particular *vocal* displays were never identified. The emphasis on how stereotyped *visual* signals are also promoted the view that *auditory* signals were stereotyped too, and the rich structural variation they often contain was long neglected. Desiring discrete signals, many scientists overlooked the grading of sound structures within the likes of bird calls; in some cases, they simply postulated that variable signals are perceived and neurologically processed as if they were discrete.

There is a biological framework by which vocal communication can be studied; it is called the logic of natural selection. Experiments in which young animals are raised in isolation from their own kind have shown that most animal signals including vocal ones are genetically controlled; they are inheritable and, in fact, inherited. This in turn means that the signals uttered by living animals are based on natural selection that, on balance, favored their ancestors. Almost everybody has heard of natural selection—the process by which evolution takes place—but few people are aware of its logic. Most beginning textbooks on evolution give short shrift to how natural selection actually works.

A frequently cited example of natural selection at work almost before our eyes is the famous case of "industrial melanism" among some British moths. Originally light in color, these moths became dark colored within a few human generations. The industrial revolution had gotten under way and soot was darkening the tree trunks on which the moths would rest. A lighter moth was more visible against darker bark, so light moths were picked off by birds more readily, few of them surviving long enough to reproduce. The light moths, in other words, were selected *against.* Meanwhile, among the preindustrial moth population were a few darker misfits, probably a rather small percentage, since they would tend to show up more clearly against the light color of the pristine, preindustrial bark. With the arrival of soot, the misfits are the ones most likely to escape predation and survive to reproduce. Thus the genes for dark moths would rapidly come to predominate in the population.

Rarely are such matters so simple. For most creatures, there are a vast number of selective pressures operating both for and against in a complex and nearly unfathomable balance. These pressures arise from the environment, the totality of the context in which the creature lives. The creature confronts all these selective pressures with the totality of its genetic heritage, including the capacity to learn from experience where that ability is present. Such are the broadly conceived dramatic elements in the evolutionary play that is acted out on the stage of ecology. It is a continuous and constantly changing narrative, in which a myriad species come and go. If for a moment we could think of natural selection as the playwright and producer of the evolutionary play, deciding what species will come on stage and which will exit via the process of extinction, we would have almost the right idea. We would be wrong in thinking that natural selection, our producer, actually causes species extinctions. Individual animals suffer the effects of natural selection, either surviving to reproduce or dying out. Extinction of a species occurs at the individual level. Natural selection works by a logic that is almost a tautology. If you are not fit, you almost surely won't survive to produce young; if you survive, the combination of characteristics by which you work out your destiny in the world must of necessity have been generally correct, at least up to the present. By surviving and reproducing, you display and embody fitness. Species—populations that are sufficiently similar to breed successfully—arise as an indirect result of natural selection. The real question is not "How do species survive?" but "How do individuals perpetuate their genetic material?" There is a biological reality to a species, but the purpose of evolution by natural selection is not to pro-

duce species. Evolution has no purpose whatsoever except to serve as the process by which individual life forms
adapt over time to the conditions of the world. Nor is
evolution interested in simplicity or complexity. Natural
selection often selects *against* complexity; consider the
tapeworm, which has lost both its mouth and gut, the
cave fishes that have lost vision, or the songbird species
with only a single song whose close relative sings twelve
different songs. The logic of natural selection does not
operate "for the good of the species." The human species is the only one that (a) has the vaguest notion that
all its members are of a kind, or (b) contains members
that pay at least lip service to the idea of acting in behalf
of the entire species, or (c) contains a handful of members that actually do act in behalf of the species.

NEPOTISM

Saints, directors of charitable organizations, philanthro-
pists, Samaritans, some doctors and nurses . . . there are
various forms of altruistic behavior in human society. In
nature there is altruism, but the word means something
quite different in the mouths of biologists. Honeybees,
though hardly the sort of creature that can decide such
matters on a daily basis, are considered altruistic because
they die after stinging in defense of their colony. That is
because it is now generally considered axiomatic that
genes are selfish. There is in genes an ineluctable tendency to seek ways to reproduce themselves in another

generation, and another, and if necessary at the expense of the gene of another of one's own kind. A territorial male tiger, for example, will do everything in its power to keep another male's genes out of the gene pool—that is, out of the tiger cubs he likes to think of as his. On the other hand, there are many examples of what appear to be Good Samaritanship among birds and mammals— young birds (like scrub jays) helping their parents take care of the nestlings, and so forth. In most such situations, these helpers are siblings or uncles or aunts, who share at least a large fraction of their genes with the parents or the young; if the helpers are unable for whatever reason to produce their own young, then helping out with close kin actually serves to perpetuate *some* of their own genetic material. Understanding this has led biologists entirely away from the idea of evolution working in behalf of species, and once shed of that concept, they have been able to make some of the more interesting insights into the behavior of animals. As for the altruistic bees: workers share 75 percent of their genes with each other, not just 50 percent as in most sexual matings. Drones are formed from unfertilized eggs (they are haploid). Since they received only one chromosome set, they can pass on only one set to the young, with the queen passing on one of the two sets she carries to each daughter, or worker, bee. Thus the daughter bees share the same chromosome set from their father, the drone, but each has one of the two possible sets from her mother, the queen. Thus all are 100 percent related from their father, but only 50 percent related from their mother's side, or an average of 75 percent related to each other

(100 + 50 = 150 divided by two parents = 75 percent related). In this context, when a daughter bee however unwittingly sacrifices herself in behalf of the colony, it makes genetic sense and, in this sense of the word, can be considered altruistic. (This is a semantic case, in fact, where it might have been better for scientists to invent a different word, since we expect far more from human cases of altruism than this thinly disguised special pleading and nepotism.)

That members of the same species, be they robins or frogs, share the same signals does not mean that they cooperate. On the contrary, it appears that the signals animals use may be utilized to *coordinate* affairs to each individual's own advantage. If, in such a situation, both the signaler and the receiver benefit mutually, it may appear to be a case of cooperation, but the actions arose for other reasons. An individual animal uses signals that penetrate through the various filters of another animal in order to affect its behavior or physiology, and it uses those signals that, over generations, have been favored by natural selection, just as the male robin's red breast has been favored as a signal of identity by natural selection. Animals emit the sound signals they do because those are the sounds that have worked up until now as one feature of a great panoply of other behaviors and actions and features and ways of life that have also worked up until now. An animal's sound signal can be thought of as having an immediate (proximate) function and a long-term (ultimate) purpose.

In the short run, a sound signal is a response to a particular situation—either internal or external or both—whereby the animal may achieve some effective change in its environment, usually vis-à-vis another animal or group of animals. The response, the signal itself, is governed by what has worked before in the animal's life and the lives of its ancestors. If that signal is no longer the best response because of a change in circumstances, then it will be selected against. This does not mean that an animal that makes one false utterance is a goner (though of course that can happen too), but that an animal making the wrong responses—in this case the wrong sound signals—over a long period of time is not well adapted to its environment in the matter of signaling, and this maladaptiveness, along with other negative factors, could well bring about its downfall. In other words, sound signals function in conjunction with other components of the animal's existence, and this overall existence is assessed by nature on a time scale that is at least as long as the animal's reproductive life, and probably longer. So the question is: How does auditory signaling by an individual affect its fitness, its ability to survive and pass along its genes to a new generation?

To return to the tail-wagging squirrel staring down a snake: The logic of natural selection forces studies of communication to encompass all the potential sources of selection on an animal, many of which are not usually considered components of animal communication under the informational perspective. To take a completely hypothetical example: The tail-wagging of the squirrel might have a *proximate* role of some kind to do with the snake, but at the same time have an ultimate role as well—to avoid surprise attack from another quarter altogether.

■ ■ ■ ■ ■

RECAPITULATION

It seems like a good point to hit the pause button, and take a look at where we are. We've looked into a number of remarkable systems of animal communication, not to show the entire range across the animal world by any means, but to point out the nature of their variety, both vocal and nonvocal. And we've considered some ways of regarding such systems. A number of questions have been raised, either directly or tangentially.

1) The question of animal language. Songs like those of the humpback whales seem to show a curious kind of syntax, with phrases and themes coming and going in what appears to be a constant framework. Is this something like a human being's wired-in propensity to use human linguistic syntax regardless of the actual language spoken? The more elaborate songs of certain birds may also show a syntactical capacity. At the same time, bird calls, and to an even greater degree the alarm calls of vervet monkeys, appear to have a nearly wordlike specificity—their calls referring to something particular in the environment. Bees can evidently indicate with great precision something as exact as distance over the ground by the intensity of their movements; they can also, by changing the angle of their movement, show

Drawing by John Anderton

Fig. 8. Ethologists A. Stanley Rand and Michael Ryan couldn't understand why South American male túngara frogs, calling to attract females to ponds, did not always put "chucks" at the end of the downslurred whine that begins the call. The "chucks" were more attractive to females than the whine alone, but also, it turned out, to the predatory bat *Trachops cirrhosus*. A male túngara frog strikes a balance between mating and longevity by giving "chucks" when other males are calling "chuck" too.

direction relative to the sun. This would seem akin to using abstract symbols to stand for human words and sentences.

More to the point, is there any point in thinking of these extraordinary talents in such terms as *language,* a word overloaded with our own unconscious and conscious understanding of what we do when we open our mouths?

2) Most attempts to listen in on animal vocalizations have been either to get a specific notion of what they are like—that is, what sounds are unique to a given species— or to find out their immediate effect on the part of the animal's environment that can hear the emitted sounds. Comparisons across species have been largely trivial: Is the whale's song more, or less, complicated than a bird's?

In studying the effects of animal communicators on their environment, scientists have tended to resort to the concept of information theory, assuming that data are encoded in an animal signal that, upon reaching the ears and brain of a listener, is decoded. This suggests a particular kind of transaction—the passing and receiving of something quite abstract called *information*—and it raises the question (without answering it) of how this could have arisen in the course of evolution. (It also implies that listeners are basically passive, like a telephone waiting to be rung.)

In other words, there is presumed to be an immediate advantage to the utterer *and* to the listener in such a transaction and, implicitly, a long-term advantage for both participants in terms of their overall fitness within the environment in which they live and in which their ancestors have evolved. Biologists think of these as proximate and ultimate situations.

3) So far, analyses of animal communication, especially that accomplished by amphibians, reptiles, birds, and mammals through vocalizations, have not helped us make much sense of these phenomena as a connected whole. It is clear, on the other hand, that animal species arise over time through a complex series of selective pressures acting in many ways on the *entire* complement of features of a given animal—its organs and its behavioral patterns, its automatic and learned responses—and on both the form and function of these features. Up to now, animal vocalizations have not been considered with any great depth in this context: The actual relation between form and function of such animal sounds has tended to be ignored in the evolutionary context in which virtually all other features of an animal are routinely considered. As a result, the similarities of animal vocalizations across species have tended to be overlooked. This is akin to a physicist trying to understand the behavior of the electron in a hydrogen atom separately from the behavior of electrons in other atoms: No general theory can emerge.

Many of these matters will arise again and some will find answers that, if not complete by any means, will at least permit the asking of more precise questions that do have a hope of being answered. But to this end we are going to ask you to keep in mind the glimmerings of a new *concept* for the voices of nature. Instead of thinking with such emphasis about the animal making the sound—the bird calling, the tiger growling, the dog barking—we suggest that the other animals out there, the ones who are listening (or not listening), are more important than they've been given credit for. *It is likely that the listener is the crucial figure in animal communication and that, over the eons, it has been the listeners that have shaped the very sounds animals emit.*

This may seem a vague and utterly unfamiliar idea, but it shouldn't be for anyone who has watched, say, a stand-up comedian in a nightclub or on stage. The comedian appears with a repertoire of rehearsed jokes, in various categories ranging from subtle to coarse. He uses a few; they do not bring the guffaws he had hoped for, so he switches to a slightly different category. Eventually, from listening to the response from beyond the footlights, he begins to get in a groove that matches the expectations (in this case the sense of humor) of his audience. He came to please (so he would get paid and be asked back) and the audience came to be entertained—two slightly different goals—and because the comedian is sensitive to his audience, he eventually fits his routine to their requirements. Thus, to a considerable extent, it is *the audience that has shaped the routine.*

With this in mind, we can proceed to something quite concrete, a general theory of certain kinds of animal vocalization. What is needed is a way of listening in on some animals, and listening in on the listeners—understanding what all the racket really signifies and what it accomplishes. We are looking for a pattern in the tower of animal Babel that fits within the overarching framework of biology, so aptly described by ecologist G. Evelyn Hutchinson as an evolutionary play in an ecological theater.

Four

||||||||||

THE GROWL-AND-WHINE
SCHOOL

*In which a diversion during an
inclement time leads to a General Theory
of vocal communication among birds
and mammals; our debt to ancient
amphibians and reptiles is made plain;
and the General Theory receives
convincing corroboration, lending
fulfilling meaning to the statement
Q.E.D., or "thus it is shown."*

Suppose you are a young biologist whose most pleasur-
able professional moments have been spent in the tropics, tracing the lives of the rain forest's songbirds. For
the purposes of making a living, however, you find yourself required to spend most of your time indoors in the
decidedly untropical surround of Maryland. Naturally, it
would occur to you to get some birds and put them in the
office or the lab next door—not only to attempt some
useful ornithological work but also to have the pleasure

of bird song while you are imprisoned. It would occur to you, perhaps, that most birds of this intemperate region in the Temperate Zone stop singing after early summer, so you would select a bird that tends to sing year-round, preferably a local one, so that if you chanced to notice something interesting about it, you could use this as an excuse to leave the office and go off to the woods and check a lab finding against real life. And if you were the senior author of this book, you would choose the Carolina wren. These birds have a social system more like that of tropical birds than of Temperate Zone bird species: They form permanent pair bonds and defend their territories year-round, which is probably why they also use all of their vocal repertoire year-round, thus playing into your hands. Or ears.

The wonderfully variable trill of the wren's song provides a pleasant background sound, but before long you begin listening in on the *chirts* emanating from the cage and realize that these vary quite a bit in tone and temporal frequency. And this is not supposed to be, according to ethological canon. It is conventional wisdom that bird calls are rather discrete, stereotyped in much the same way as an excited bird's display of feathers can be. For example, a raven gone a-courting will put on a very specific, exaggerated, and stereotyped act before a female. He will bend forward, head down, beak open, wings slightly raised, and make a distinct, silent coughing motion with his head. He looks like a caricature of a raven vomiting. This is no accident: The pose almost surely evolved from the act of coughing up a piece of food to place at milady's feet, thus demonstrating a willingness

to take good care of her. Over generations of early ravens (or protoravens, more likely), this actual act of feeding became somewhat abstracted—symbolic, you might
say—and exaggerated so as to stand out visually from the
confusing background noise of daily raven behavior: In
short, a very distinct and discrete act signaling a particular intent or state of mind.

Since it was such nonvocal signals and displays that
first attracted the attention and analysis of bird ethologists (those who study the evolution of bird behavior, as
opposed to the evolution of bird anatomy) assumptions
about bird displays tended to be carried over into studies
of bird vocalization. While the songs of some birds, such
as the mockingbird, can vary enormously, bird calls like
the wren's *chirts* were simply assumed to be uniform,
discrete, and stereotyped. A Carolina wren's *chirt* was a
chirt was a *chirt*.

(In truth, there had been some suggestions that the
physical structure of bird calls was not arbitrary. It has
been pointed out that some sounds, like certain alarm
calls, are designed to provide clues to the sender's location while others are not (see Chapter Two). But the
variations in the Carolina wren's *chirts* seemed to have
some other function besides expressing location.

▌▐▌▐▌▐▌

THE VARIOUSNESS OF *CHIRTS*

The rigors of scientific experimentation were needed,
and the senior author undertook just that in 1976. A

captive Carolina female was placed in a cage in a woodlot near College Park, Maryland, and alongside her a tape recorder that played the male song. To this, the female responded with her typical *chatter* vocalization, and for wild Carolina wrens in the neighborhood this all sounded like the work of a new wren pair that had showed up and was muscling in on their territories. In turn, the wild pairs would approach, prepared to repel the invaders, only to be trapped in a mist net, a fine-meshed net like those found on badminton courts but much larger, that is stretched across a bird's usual route. Eight pairs were thus duped, and placed in hardware-cloth cages in a room completely lined with white blankets to reduce echoes and create a uniform background. A hawk was also placed in the room, and trained to fly from a perch that the wrens could not see right over their cages to a perch at the other end of the room. The birds could see the hawk pass overhead and they could see it sitting on the far perch. Their vocalizations were tape-recorded and the birds were observed through a one-way mirror, their behavior clocked so that it could later be matched to the tape recordings.

It was soon clear that the *chirts* served in part as a predator surveillance system; the *chirts* increased in tempo when the hawk flew overhead or, when sitting on its perch, it moved its head, or especially when it glared directly at the wrens. It was chiefly the females that *chirted*—males sounded the alert only about 15 percent of the time, and only after the females had sounded especially alarmed. This pattern is explainable by the fact, observed in nature, that lone females in this species

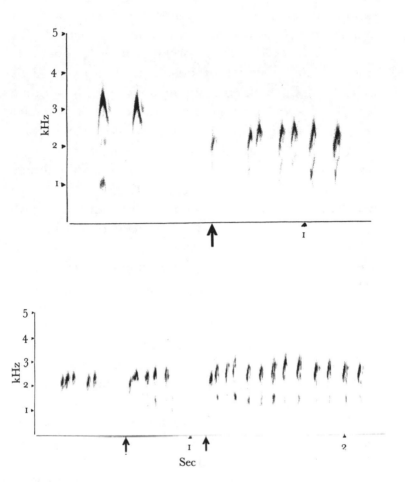

Fig. 9. Carolina wrens react to a hawk. Above, two upside-down chevron calls, or *barks* (see Fig. 1), are followed by seven *chirts* when the hawk moved (arrow). Below, a series of *chirts* with two hawk movements (arrows). The *chirts* remain at a higher pitch after a hawk movement than when the hawk is still.

cannot maintain a territory against the superior power of a mated pair. They get driven off and eventually die for lack of food resources. On the other hand, a lone male can and does successfully defend a territory. So it is crucial that a female form a pair bond quickly, within a few short weeks after fledging; thereafter it is to her advantage to take the risk of becoming more easily located by a predator and to sound alarms in order to protect her mate from predation, even to let him forage in the nearby presence of a predator. The alternative—to be a widow, particularly in winter when pair bonds have long since formed and territories been established—is calamitous.

In any event, both female and male *chirts* seemed to be sounds of attention, of alertness. When the hawk flew overhead, these sounds became more high-pitched and frequent, but when the hawk was sitting still on the perch they were lower and harsher. When the perched hawk turned to look at the wrens, the female would often emit an even lower, less tonal *dit,* a sound used in nature to repel another female from a territory, or when mobbing an intruding predator. Another variation noted was that the female's *chirts* would be emitted in variable sequences of one at a time, two, three, four, or five, with changes in sequence (and/or changes in pitch) alerting the male that the hawk was moving, at which point the male wren might sound alarm calls too. Otherwise, the males would keep on foraging or preening while the females kept up a low-intensity alert.

Field studies out in the woods followed, with a captive hawk being placed in wren territories and their reactions observed, their vocalizations taped. The *chirts* were found to have distinct characteristics or structures, depending on circumstances. When a wren flew away from the hawk, escaping this fearful presence, it emitted a high, shrill *chirt,* but when it seemed to be under attack, the *chirts* were both harsh and shrill at the same time, suggesting an ambivalent state of mind; the threatened wren didn't know if it wanted to attack (a harsh *chirt* is used in territorial defense among wrens) or to flee. In other words, in addition to the tempo of alarm calls having a meaningful effect between members of a pair, the calls themselves also showed measurable gradations of structure—of acoustical form—that in turn appeared to reflect quite subtle gradations in the bird's state of mind, its mood, or what could be thought of as its motivation of the moment. The form of the *chirts* could of course be seen clearly on spectrograms made from the tape recordings. A *chirt* indicating an ambiguous state of mixed aggression and fear was in the shape of an inverted *V* or a chevron, showing a rise and fall in pitch. The escape *chirt,* on the other hand, was a rising shape. Higher pitch seemed to mean fear, while a lower pitch seemed to mean aggression. The actual form of the sound—its acoustical shape—was evidently tied to the mood or motivation of the vocalizer and permitted a listener to read that mood and take whatever action seemed to be called for, if any. Form, in Carolina wren *chirts,* was thus directly tied to function.

█ █ █ █ █ █

A GENERAL THEORY

Now this may not seem as electrifying as the discovery of the structure of the atom or the nature of the electromagnetic force, but it turned out to be more important than one might think a wren's *chirts* could be. It soon led to what could be called a General Theory of one kind of animal communication.

The graded, qualitative differences in the supposedly stereotyped alarm calls of Carolina wrens seemed important, and it wasn't long before the sounds emitted by a large range of animals in close conversational proximity to one another were reexamined in this light, using among other modes of analysis great blizzards of spectrograms, which in some cases could conveniently be "ground-truthed" by wandering around the National Zoo. (Spectrograph machines, it will be remembered, produce a two-dimensional drawing of a sound or sounds, the horizontal dimension indicating duration in time, the vertical dimension showing pitch. A more precise scale that scientists use is measured in kilohertz [kHz], which refers to the number of cycles per second of soundwaves.) What quickly emerged from looking at the shapes of a variety of animal sounds was startling. The wren's *chirt* occurs quite high, in the range of two to six kHz, while the bark of a dog occurs in the range of 0.1 to 0.8 kHz. But once that vertical difference between the two on the scale has been adjusted for, looked at side by side, so to speak, they appear very much alike:

█ █

a chevron. In effect, wrens bark and dogs *chirt:* It's the same thing. Both go up and down to varying degrees.

There is an old saying that barking dogs don't bite, and in fact this tends to be true. The reason lies in the chevron shape of a bark on a spectrogram. It goes both up and down, fear and aggression mixed, a sound that in effect says, "I'm alert to something here and I'm not sure how I'm going to act." When the bark descends down the scale, finally becoming a low roaring growl, it is time to get out of the way, or pick up a rock. (The authors should confess to mixed feelings about dogs: Some dogs are nice, some wonderful, and the authors like animals in general. But taken as a whole, dogs, even nice ones, and even more so, house cats, are biologically speaking subsidized predators, often wreaking havoc in local ecosystems. The authors have mixed feelings about dogs, and our immediate response is a mental bark.)

When a "General Theory" based on these findings was first announced in the scientific literature in 1977, the title of the article noted their occurrence in "some mammal and bird sounds," and included a list of twenty-eight birds and twenty-eight mammals along with descriptive or phonetic notations of the sounds they use in hostile, friendly, or appeasing contexts. There is of course some danger in using a human observer's description or phonetic rendering of a sound. For example, in studies of ravens, very loquacious birds, what one biologist may hear and write down as *cawlup* may be represented by another biologist as *kowah.* It could be two distinct sounds with different functions, or merely a

Fig. 10

Mammalian Sounds Used in Hostile or "Friendly," Appeasing Contexts

Species (family)	Hostile	Friendly or Appeasing
Virginia Opossum, *Didelphis marsupialis* (Didelphidae)	Growl*	Screech
Tasmanian Devil, *Sarcophilus harrisii* (Dasyuridae)	Growl	Whine
Wombat, *Vombatus lasiorhinus* (Phascolomidae)	Deep growl	quer-quer-quer
Guinea Pig, *Cavia porcellus* (Caviidae)	Grunt, snort	Squeak, *wheet*
Mara, *Dolichotis patagonum* (Caviidae)	Low grunts	Inflected *wheet*
Curu curo, *Spalacopus cyanus* (Octodontidae)	Growl	Short squeaks
Degu, *Octodon degus* (Octodontidae)	Growl	Inflected squeak
Spiny Rat, *Proechimys semispinosus* (Echimyidae)	Growl	Twitter, whimper
Agouti, *Dasyprocta punctata* (Dasyproctidae)	Growl, grunt	Squeak, *creak-squeak*
Pocket Mouse, *Heteromys* (2 sp.) (Heteromyidae)	Low scratchy growl	Whining squeal
Pocket Mouse, *Liomys pictus* (Heteromyidae)	Low scratchy growl	Whining squeal
Desert Pocket Mouse, *Perognathus* (4 sp.) (Heteromyidae)	Low scratchy growl	Whining squeal
Kangaroo Rat, *Microdipodops pallidus* (Heteromyidae)	Low scratchy growl	Whining squeal
Kangaroo Rat, *Dipodomys* (6 sp.) (Heteromyidae)	Low scratchy growl	Whining squeal
Lemming, *Dicrostonyx groenlandicus* (Cricetidae)	Snarl, grind	Whine, peeps, squeals
Unita Ground Squirrel, *Citellus armatus* (Sciuridae)	Growl	Squeal
Maned Wolf, *Chrysocyon brachyurus* (Canidae)	Growl	Whine
Bush Dog, *Speothos venaticus* (Canidae)	Buzzing growl	Squeal
Coati, *Nasua narica* (Procyonidae)	Growl	Squeal
Large Spotted Genet, *Genetta tigrina* (Viverridae)	Growl-hiss	Whine or groan
African Elephant, *Loxodonta africana* (Elephantidae)	Roaring, rumbling sounds	High frequency sounds

*Verbal or onomatopoetic (italics) renditions of sounds quoted from source author's descriptions.

Fig. 10 (Continued)

Species (family)	Hostile	Friendly or Appeasing
Indian Rhinoceros, *Rhinoceros unicornis* (Rhinocerotidae)	Roaring, rumbling	Whistling
Pig, *Sus scrofa* (Suidae)	Growl	Squeal
Llama, *Lama guanacoe* (Camelidae)	Growl	Bleat (long distance only)
Muntjac, *Muntiacus muntjac* (Cervidae)	Not given	Squeak
Squirrel Monkey, *Saimiri sciureus* (Cebidae)	Shriek calls, *err*	Peep calls, trills
Spider Monkey, *Ateles geoffroyi* (Cebidae)	Growl, roar, cough	*Tee tee*, chirps, twitter squeak
Rhesus Monkey, *Macaca mulatta* (Cercopithecidae)	Roar, growl	Screech, clear calls, squeak, nasal grunting whine, long growl

Avian Sounds Used in Hostile or "Friendly," Appeasing Contexts

Species (family)	Hostile	Friendly or appeasing
White Pelican, *Pelicanus erythro-hyrchus* (Pelicanidae)	Harsh nasal growls	Not given
Mallard, *Anas platyrhynchos* (Anatidae)	Loud harsh *gaeck* (♀)	Soft whimpers: *kn* and *q*
Sparrow Hawk, *Falco sparverius* (Falconidae)	Harsh *chitter*	Whine
Bobwhite, *Colinus virginianus* (Phasianidae)	Loud, rasping "caterwauling"	*Tseep; squee*
Ring-necked Pheasant, *Phasianus colchicus* (Phasianidae)	Hoarse *krrrrah*	Squeak (♀)
Solitary Sandpiper, *Tringa solitaria* (Scolopacidae)	Harsh, metallic sound	Rising shrill whistle
Stilt Sandpiper, *Micropalama himantopus* (Scolopacidae)	*Trrr*	*Toi, weet*
Cassin Auklet, *Ptychoramphus aleutica* (Alcidae)	Growled *krrr krrr*	*Kreek*

Fig. 10 (Continued)

Species (family)	Hostile	Friendly or appeasing
Orange-chinned Parakeet, *Brotogeris jugularis* (Psittacidae)	*rrrr*	Low-intensity "chirp"
Burrowing Owl, *Speotyto cunicularia* (Strigidae)	rasp	*eep*
Red-headed Woodpecker, *Melanerpes erythrocephalus* (Picidae)	Chatter, rasp	Not given
Harlequin Antbird, *Rhegmatorhina berlepschi* (Formicariidae)	Growling *chauhh*	*chee*
Chestnut-backed Antbird, *Myrmeciza exsul* (Formicariidae)	Snarling nasal *chiangh*	Musical chirps: *cheup*
Eastern Kingbird, *Tyrannus tyrannus* (Tyrannidae)	Harsh *zeer*	High-pitched *tee*
Barn Swallow, *Hirundo rustica* (Hirundinidae)	Deep harsh stutter	Whine call
Purple Martin, *Progne subis* (Hirundinidae)	*zwrack*	*sweet*
Mexican Jay, *Aphelocoma ultramarina* (Corvidae)	Not given	Variable *weet*
Scrub Jay, *A. coerulescens* (Corvidae)	Harsh rattle	*wheu, scree*
Dwarf Jay, *A. nana* (Corvidae)	Harsh rasp	*shreeup*
Common Crow, *Corvus brachyrhynchos* (Corvidae)	Growl	Soft and plaintive
Carolina Chickadee, *Parus carolinensis* (Paridae)	Click-rasp	Lisping *tee*, soft *dee*, high *see*
Blue-gray Gnatcatcher, *Polioptila caerulea* (Sylviidae)	*peew*	*spee*
American Redstart, *Setophaga ruticilla* (Parulidae)	Snarl	*zeeep*, high pitched *titi*
Yellow-headed Blackbird, *Xanthocephalus xanthocephalus* (Icteridae)	Harsh, nasal *rahh-rahh*	*pree pree pree*
Crimson-backed tanager, *Rhamphocelus dimidiatus* (Thraupidae)	Rasping harsh hoarse notes	*Seeeeeeeet*
Brown Towhee, *Pipilo fuscus* (Fringillidae)	Snarling throaty notes	*Seeep*, squeal duet
Common Redpoll, *Acanthis flammea* (Fringillidae)	Harsh *cheh cheh cheh*	*sweeee*
African Village Weaverbird, *Ploceus cucullatus* (Ploceidae)	Harsh growl	look!; high squeal

dialect variation of the same sound. In any event, even with this difficulty as a caveat, the lists showed a most persuasive coherence, suggesting that a general set of rules existed that seemed potentially applicable to all birds and mammals. Because the structure of the sound appeared to be firmly tied to the motivation of the vocalizer, these rules were called Motivational/Structural Rules, or M-S rules, and if they were indeed applicable across the board, they not only permitted listening in on an animal's state of mind, but also opened a door to how at least a great deal of animal vocal communication may have evolved.

█▌█▌█▌█

THE IMPORTANCE OF BEING BIG

Darwin had noted the principle of *antithesis* in the nonvocal expression of emotions in animals. An aggressive animal like a dog will raise the hair on its neck and shoulders, appearing larger to an adversary. On the other hand, a fearful dog, seeking to avoid trouble, cowers; it lowers its head and slinks off close to the ground, appearing smaller. Size, then, is a factor—even apparent size, it turns out. But actual size is of major importance in such matters as a fight over territory or females. The bigger male usually wins out, and this tends to be true from frogs to elephants. When a big male elephant in *musth* comes barging into a female herd, responding to a female's infrasonic siren calls, the younger males who may have gathered there for the same purpose generally

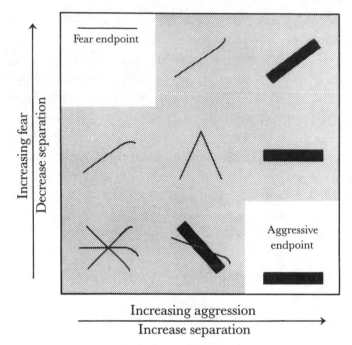

Fig. 11. A diagrammatic representation of sound structures to illustrate the Motivation-Structural Rules code. Nine hypothetical sound spectrograms vary from a thin line (tonal sound) placed high in the block (high pitched) in the upper left to a broad band (harsh, growl) low in its block (low pitched) in the lower right corner. Sounds become harsher from left to right and from top to bottom and more tonal from right to left and from bottom to top. In the upper right is a sound combining fear and aggression: Often called a scream, it is rising in pitch yet harsh due to the combination. In the middle is an upside-down chevron called a bark, completely in the middle and, therefore, not denoting either endpoint. For example, a dog whines when friendly, growls when aggressive, screams when in pain during a fight (but fighting back because it is unable to flee without risking a bite), and barks when indecisive but stimulated by, for example, a car door being shut.

get out of the way, saving their energy and their hides. In nature, most fights are avoided, the bigger animal's threat posture—or the threat posture in combination with a low (for that species) growl—often being enough to persuade the smaller one that the game is already up.

Evolution then would seem to favor larger size within a species. There are limits to this, as in any aspect of biological design, which is always at best a compromise. The male lion's mane makes him seems larger than he is. An even larger mane might be advantageous, unless it was so large he tripped over it. Another compromise: The bright blue of a whiptail lizard's tail evidently makes him more attractive to females, but also more visible to the local roadrunner. On balance, over generations of natural selection in whiptailed lizards, the need to attract females seems to have taken precedence over the danger of predation—particularly inasmuch as the lizard is extremely quick, and that its tail, if caught by the roadrunner, will fall off and regrow. In fact, at close quarters (when being grabbed, for instance), the bright blue may tend to direct the roadrunner's attention to the tail rather than a more vulnerable spot. Natural selection doesn't create perfection (whatever that might be) but a momentary optimization of a range of features that go to make up an organism and permit it to continue. Larger size is a feature that, within reasonable species limits, confers an advantage, as does the appearance of greater size, as accomplished, for example, by raising the hackles. A small mammal that can suddenly appear larger in this way may persuade a marauding predatory bird that it is too big to be taken. So large size, or apparent large size, works not only within a species but between species in some cases.

There are other ways to indicate large size, notably by sound. No matter what you do to a snare drum, you cannot make it emit a sound as low as that of the larger bass drum, and that principle may well be why evolution has favored the low growl as a sign of aggression. Every bird or mammal that attempts to make another animal move away, it turns out, uses a low-pitched, usually harsh sound—a growl . . . even humans. We can tell a pest to go away, using the arbitrary sounds we have invented that add up to words, such as "Go away!" "vamoose!" or "vah yah sey" (phonetically in Spanish). It may work, but if we are serious we say it with a lower than normal pitch, usually trying to get the last word or syllable lower in pitch:

Go away!
simply is not as convincing as
Go a-way!

The intonation adds to the meaning and, ideally, we reap the benefit: The pest leaves. Even a nonverbal grumble is often enough to achieve the desired result, regardless of what arbitrary human language the pest may speak and understand.

There have been, as pointed out, some students of animal communication who considered the sounds animals emit as vocalizations to be arbitrary, like our human speech (though not our growls)—in a sense, species artifacts. But if such sounds were indeed arbitrary, merely accidental by-products of whatever has led to the evolution of a given species, then surely somewhere in the animal kingdom we would find some other vocalization besides the growl used by highly aggressive animals. So far, we do not.

Why the universal growl? Like a bass drum as opposed to a snare drum, a large voice box produces a lower sound. In physics this is explained as follows: The larger the object, the lower its resonant frequency, the longer the wavelength of sound it produces, and the lower the perceived sound. It seems reasonable to think that evolution favors the growl in aggressive animals because a low sound makes the growler seem larger to a rival, and thus all the more threatening. Evolution, as it were, presented mammals and birds with this signal on a platter, because the size and voice-pitch relationship already existed in the amphibians and then more recently the reptiles, from which both birds and mammals arose.

It is more than a curiosity that amphibians and reptiles continue to grow larger and larger until they die, whereas most birds and mammals stop increasing in body size once they reach sexual maturity—in humans it can take several years. There are a few exceptions to this in the mammalian world: Elephants continue to grow larger throughout life, as do *all* amphibian and reptiles.

There are, of course, many things that can have retarding effects on the size of an individual frog at a given age—lack of enough food, for example—but for the most part, the older the frog, the bigger it is, and the bigger its body and voice box, and thus the lower its resonant frequency, the longer its sound wavelengths and the lower its perceived voice. Furthermore, typically, when the evening chorus of frog calls (messages calling in the females) start, females tend to head for the biggest frogs, a feature they know only from the low pitch of its call. Age, as revealed by size, in turn revealed by sound, suggests superior or at least well-tested survival skills. In some frog species, a smaller male will sidle up to a larger, lower-voiced boomer and wait silently in the dark to grab any female that goes by on her way to the bigger frog. The same principle seems to work in those vocal reptiles, alligators and crocodiles, whose roars are semaphores of size, age, and, presumably, staying power in a dangerous world.

So size is symbolized by sound in these ever-growing animals. For some female frogs, lab experiments have shown that the voice alone is more compelling than actual vision. Furthermore, cheating is impossible, at least theoretically, because even if a small frog developed an outsized resonating throat pouch, it would still probably wind up having to fight with a larger male at some point. Such contests would tend to maintain truth in advertising. This sound/size symbolism is probably the ancestral condition of vocal communication among birds and mammals.

The restless process of evolution was not to leave this condition at a status quo, any more than it would leave the world entirely to the dinosaurs and reptiles. Just as the inner ear of terrestrial vertebrates gradually evolved from the primitive gill arches of fish, natural selection shaped something new in vocal communication out of the herpetological condition. Sound came to symbolize more than size: Among animals that can be said to have the equipment to experience a wider range of what we call, in ourselves, emotions, it included mood.

■I■I■I■

SEEMING BIG (OR SMALL)

The relentless evolutionist asks: What advantages might have been conferred by this recasting of size/sound symbolism into mood/sound symbolism? It boils down to avoiding energy-sapping battles. If competition within species tends to promote greater size, there are limits to growth, as it were—energetic ones. If the diameter of a sphere is squared, its volume and thus its weight is cubed. An elephant two times taller and longer than normal would be *six* times heavier. To maintain such a vastly larger "plant" requires vastly more fuel, not to mention a great deal more investment in the likes of vastly stronger legs. At some point, different for every species, the law of diminishing returns comes into play: An additional increase in size isn't worth the added investment and operating costs. (Another way to think about this is the lighting of a candle in a dark room. The amount of apparent light is great. A second candle adds to it significantly, but the thirtieth candle barely adds enough light to be noticed.) Large animals that fight among themselves for resources can generally get on well enough as a species so long as the environment is salubrious, cooperative, and provides ample resources. But an environmental shift of any drastic sort can render such large creatures vulnerable: They may not be of an optimal size energetically for the new world that confronts them.

The argument is that genes that promote fighting in animals where large size is an advantage (though a costly one) will tend to be replaced by genes that promote the much less costly business of communicating one's aggressiveness vocally—*if* such communication turns out to be an equally effective means of obtaining resources from the environment or in holding on to them. Animals able to express their intentions, and to read their fellows' intentions from such communication, would gain an advantage, one that put less of a premium on sheer size and more on a larger or more complex brain, one with a more refined memory by which to learn from (painful) experience, for example. And indeed birds and mammals have an additional brain structure added on to the reptilian brain—the mammalian neocortex. In birds, an equivalent structure is called the hyperstriatum. The capacity to read more accurately the state of mind of one's fellows goes hand in hand with a more intricate social behavior. And in the dynamic relationship between brain size or complexity, social relations, vocalization, and body size, one can see that it was almost inevitable that the low, harsh growl, so suggestive of great size and power, would become the adaptive sound structure in hostile situations. There had been, after all, millions of years during which animals actively listened and perceived low sounds as dangerous.

How then did the whine—the high tonal sound—come about as an indication of fear or the related intention to appease? From the mouths of babes, which are inevitably smaller than those of their parents and, being small, emit higher-pitched sounds in *any* context. Small things are generally less threatening than big things. So when the size/sound symbolism of amphibians became a motivation indicator in the higher vertebrates, the sound made by small creatures became a sound of appeasement, or even fear. It makes perfect sense that an adult should use a high-pitched sound when it is trying to reduce aggression in another adult (by summoning up the other adult's parental urges) and/or tries to eliminate the other adult's fearfulness by seeming small.

Thus can the end points of the code—the growl and the whine—be explained in evolutionary terms. We have seen earlier that somewhere midway along the continuum from high to low pitch is the bark, rising toward the whine and then falling off rapidly—a mixture of fear and aggression that seems to signify alertness and indecision. In this light, consider the motivations of a goose who finds a fox approaching her as she guards her nestlings or eggs. She is both hostile and fearful at the same time. And both emotions or states of mind could appear in whatever vocal sound she emits. It simply won't do, even as she raises her wings in a threatening posture, for a squeal of fear to accompany a low aggressive sound. So she doesn't vocalize at all. Instead, she hisses—a nonvocal sound that betrays no fear and in fact, over time, has come to be selected as the goose's aggressive signal.

It often happens that a flock of seagulls gathers at a beach picnic, looking for handouts. If a picnicker throws a piece of bread into the flock, the bird that darts in fast enough to catch the morsel will often emit a high-pitched cry similar to that gulls emit when they are under attack by a predator or larger thief. It might seem that this is a case of deliberate and almost dishonest manipulation of the sound-symbol system, for competition over food usually elicits aggressive sounds and this evident alarm call usually has the effect of making the other birds hesitate slightly, and look about for the predator, allowing the "duplicitous" bird to get the jump on them and seize the morsel. It seems far more likely, however, that the bird that emits the "alarm" cry is honestly expressing its state of mind at the time: fear, specifically fear of all the irritable compatriots with which it is surrounded. And with a wonderful lack of awareness, those grouchy compatriots assume that the crier is scared of some foreign interloper.

PROVING THE HYPOTHESIS

The M-S rules were proposed originally as a hypothesis, an idea against which other observations and recordings of animal vocalizations could be tested to see, for one thing, if the hypothesis was universal among mammals and birds. Additional studies have tended to show that it is. Two Rhode Island zoologists directly tested the hypothesis on fifty species of mammals, ranging from marsupials and insectivores to carnivores like giant pandas and primates like gorillas. All aggressive sounds, they found, that are emitted by these animals are "low-frequency and wide band width"—that is, harsh growls—"and thus support M-S rule predictions. Fearful/friendly sounds show a trend toward conforming [to the M-S hypothesis] . . ." Here they did find a "considerable" amount of variation, possibly due to the fact that fear and friendliness are quite different motivations, possibly due to the failure of a spectrogram to measure all the parameters of a sound, and possibly also because the particular fear/friendly sounds they analyzed did not represent the true end point on the continuum for some of the animals they studied, but instead were less motivationally "pure." Friendly sounds made by large carnivores had been found, for example, to be low in frequency, "noisy" (meaning they comprise more than one tone), and repetitious. It had also been found that many mammals—baby mice, for example—emit a steady sequence of pulsed squeaks. When they are excited either positively or negatively, the rate of repetition rises, making the repetition rate an important aspect of the intensity of motivation. While the study suggested that much remained to be learned and that possibly other factors were at work, including the rapidity of sound emission, it also found the M-S rules to exist as a general pattern.

And that is pretty good for any hypothesis. Not only did it seem to prove out generally, but people were now talking about general rules of animal communication, as opposed to the fragmented species-by-species cataloguing that had long prevailed. Animal communication was on the way to becoming understood in the context of biological evolution. What is even better is that the M-S rules hypothesis provides a benchmark against which to analyze animal vocalizations, at least of a certain sort, and a viewpoint from which to try and make sense of the myriad variations that lie between the extreme end points of aggression and fear. One could now look at the gradations between those end points, the many combinations of sound structures that are possible along the continuum, and explore their momentary (proximate) as well as long-term (ultimate) effects, in addition to analyzing them in the light of such considerations as the ecological and social setting of the animal. Such an approach might well open a window on the uniqueness of "personality" of individual animals. For through the subtle variations of the sounds they emit, they reveal their state of mind.

In science, a hypothesis that *seems* to explain observed facts or events is all very well, but it only gains scientific power if it suggests actual tests, if it permits one to make predictions about facts or events as yet unknown or untested. For example, chickadees utter *chip* sounds that are termed "contact notes" when the birds are foraging in flocks in winter. The name "contact note" describes its function—keeping the flock together. The flock shares information about each member's whereabouts, thus staying close together, with the benefit that a predator cannot easily approach a flock undetected. But this explanation is really a set of untested assumptions, and it fails to provide a testable hypothesis. On the other hand, the *chip,* when examined spectrographically, is seen in a significant shape: as a chevron, the "bark" that across the board suggests something of interest has been perceived. A new interpretation is now possible. The chickadee *chips* when it spots food, or when it moves rapidly from perch to perch. The *chipping* chickadee gains if, as a result of his *chipping,* the rest of the flock is attracted to him, providing a form of protection. For the perceiver-chickadees, there are two possible advantages in heeding the message—finding food faster and being protected from predation—and these outweigh the possible disadvantage of losing foraging time while following some chickadee that is merely hopping from one foodless perch to another. Such an explanation, taking into account the structure of the *chip* sound and the fitness benefits to both sender and perceiver, suggests a testable situation: Are the *chips* as frequent or as effective in times of low food resources as in times of plenty?

And also, importantly, this approach places the communication in an ecological context: The richness or poverty of the food supply becomes important in understanding the receiver's positive response to the "contact notes."

Some of the more obvious predictions from the M-S rules concern end points on the scale: High-frequency tonal sounds indicate that the sender will not be hostile if approached; conversely, if the receiver of a harsh noisy sound approaches the sender or remains where it is, it is likely to be attacked. The more interesting predictions have to do with the gradations in between the end points. The M-S rules predict that a species that is more generally aggressive among its own members will tend to have a vocal repertoire that is harsher than a species that often lives in flocks or joins with other species in mixed groups. The latter will have a repertoire given to higher-frequency, more tonal sounds.

The eastern kingbird, a natty black-and-white fly-catcher, is a highly aggressive bird, as anyone in eastern North America knows who has watched one take offense at the proximity of an intruder, such as a crow or hawk or even a cow, and "mob" it, diving at it, emitting harsh cries, and in due course driving it off. In fact, kingbirds are highly aggressive among themselves, squabbling over boundaries, jealously guarding their mates, and hunting down insect prey. When attacked by another, a kingbird may lose the initiative and emit a high-pitched *tee,* but for the most part, its vocalizations range from harsh to harsher—essentially seven calls, each grading into another. This is, at least, the case during breeding season when this migratory bird is in the insect-rich climes of North America. But come late summer there is a Jekyll-and-Hyde transformation. As the kingbirds prepare for their migration south to Amazonia, they lose their aggressiveness and change their diets. They begin to congregate in small flocks and forage for berries and fruit. In the Brazilian forests, they follow the ripening fruit in large flocks, operating on the fringes left by more aggressive birds, responding to an attack not by counter-attack but by melting meekly into the crowd. And, inter-estingly enough from the standpoint of M-S predictions, during this period of nonaggressive flocking, the king-bird makes virtually no vocalizations. Natural selection has favored a harsh sound system for kingbirds to use during the rigors of the breeding season, a system that would not work during the rest of the year, when a peaceable togetherness is in order. For kingbirds, it seems, if you're not in a position to say something awful then it's better not to say anything at all. In any event, kingbirds would appear to be rather dramatic examples of an M-S prediction working out, particularly when compared to, say, cedar waxwings—birds that typically travel in flocks and typically emit sounds in the tonal range, as opposed to the scratchy low range.

Another interesting prediction from the M-S rules concerns the vocalization that occurs when both fear and aggression are simultaneously expressed. Here the sound rises in pitch yet maintains the harsh sound of aggression. Our human version of this is called a shriek.

▮▮▮▮▮▮
SOCIETY AND ITS EFFECTS

Yet another prediction arising from the M-S rules is this: The greater a species's social complexity, the more complete the range of sound qualities—that is, the more points along the motivational gradient—that will be expressed. In other words, increasingly complex social life will lead to increasingly subtle expression of an animal's mood.

Scientists are human, and there is always the possibility of finding what one is looking for, especially in a realm as complex as animal communication and so subject to interpretation. Someone looking for M-S rules in a given species may well find them, because that is what he seeks, however objectively he sets about making his analysis. So there was particular importance to the discovery made in the study of a species that lives in a complex society, a study made without any apparent prior knowledge of the M-S rules hypothesis, in this instance a study of squirrel monkeys by Uwe Jürgens, a German researcher, in 1978.

Squirrel monkeys, which live in bands of up to a hundred, are diminutive dwellers of the South American rain forest. Greenish above, yellow below, with white "spectacles" on black faces, they are highly popular zoo specimens. Jürgens's study was done on captive members of this species. In all, he found that these monkeys have a repertoire of some fifty identifiable sounds, all of which could only be explained as indicating one or another gradation of several emotional states. Jürgens was quick to point out that there is no reason to expect that the monkeys experience these emotional states in the same manner as a human would. Like all of us peering into the opaque membrane that separates us and our elaborate language from the rest of the animal world, Jürgens had to fall back on descriptive analogy.

For example, squirrel monkeys emit a variety of low, atonal sounds that Jürgens felt had to be related because of their acoustic structure, even though they varied from a contented purr (as when a young animal is being suckled, or when a group is huddling) to an aggressive growl emitted when, for example, a strange monkey arrives on the scene. Even more aggressive is a spitting sound. Jürgens considered all these to be sounds of self-assertiveness, each expressing a different degree of excitement. The wider the frequency range—that is, the wider the spectrographic drawing of the sound—the higher the probability that the animal will attack. (Similar purrs, rasps, and growls have been found among many other primate groups, including macaques and gorillas.) At the other end of the continuum is a group of sounds ranging from chirps to peeps to squeals, all of which express varying degrees of "social unease, or lack of confidence." Most chirps and peeps apparently served to keep others in the group in contact with the caller. Chirps were apparently used to draw the attention of others to the chirper in a nonaggressive manner. Certain long and short peeps are used typically during the exploration of the environment when the animals are spatially separated but not out of sight of one another: A different sound, called the isolation peep, is uttered when the sender loses sight of the others. Such calls are usually returned in kind. Also, peeps are emitted alternately by animals at play—an ascending peep—which can slide into a more shrieklike sound if the playing gets too rough. Squealing and yelling are the most excited calls of this associated group: Upon seeing a snake, a squirrel monkey may yell loudly, as will a male who is out of sight of the band when some of its members are noisily copulating.

So while each of these two groups of sound (each part of a fairly distinct continuum of acoustic structure) can signify varying degrees of intensity, of self-confidence or a lack of it, each group itself can be positioned on a larger emotional/acoustical continuum. And, in between, as it were, are other associated groups of sounds. The twittering-chattering-cackling group consists of contagious sounds that have what Jürgens calls "positively reinforc-ing value," in a sense recruiting others. *Twittering* sug-gests pleasure during feeding or the expectation of feeding. *Chattering* is heard during feeding after a long food deprivation period. *Cackling* appears to suggest a component of "bellicosity"—a relatively self-confident call to others to share in a relatively low-threat situation. Yet another group is called chuck-yapping-alarm peep: The *chuck* is emitted by huddling groups that are becom-ing restless, by mothers calling infants to them, and in minor disagreements over food. *Yapping* is a typical mob-bing call, directing the group's attention to a potential predator—it is uttered in chorus and only after safety has been achieved, never during flight. The *alarm peep* is a typical high-pitched alarm call, followed immediately by flight. The sudden appearance of a large bird or snake brings forth the alarm peep. Jürgens suggested that all three calls suggest agitation, which is to say, varying degrees of concern but all with a low component of ag-gression. The *chuck,* for example, which appears nearly neutral, expresses a very minor annoyance and very low aggression.

In general, Jürgens found that lower-pitched calls meant a higher degree of "aversiveness" on the part of the caller, but in some instances, depending on the group of sounds in question and the context in which they were emitted, a higher tonal sound also indicated "aversiveness." He suggested that there might be a cor-relation between the intensity of emotion expressed and the tension of the sound-producing muscles. In any event, Jürgens's study (which is matched in kind by studies of other primate vocalizations) shows considerable subtlety in expressing a range of mood and context, just what one might expect in a group of animals that lives in large bands and whose livelihood depends on a constant mediation of the delicate and always shifting balance between the needs of the individual and the functioning of a group, a situation familiar to all humans as well.

What is also interesting in this study is that, generally speaking, those sound/symbols of the squirrel monkeys that seem to stray outside the more precise conditions predicted by the M-S rules are mostly longer-distance communications. M-S rules apply to short-distance communication, which is acoustically a different matter. This same phenomenon, along with the predicted difference between social and solitary animals' vocal repertoires, is to be seen perhaps in even bolder relief in a study of three canine species of South America.

None of the three is especially well-known to the public, even the South American public, but each represents a different approach to canine life. Like virtually all canines—wolves, foxes, jackals, and what are called wild dogs—the three South American canines have in common the classic growl-and-whine vocalizations. These, in their simple form, are no doubt the evolutionary base upon which each has wrought its own variation, depending on the requirements of its social and ecological niche.

The maned wolf is a strangely wonderful looking animal with large erect ears like a fox and long thin legs. It inhabits the lowland woods and open plains of central

South America, in Brazil, Paraguay, and Argentina. An omnivore, it hunts deer and any smaller form of life, even rooting in the dirt for snails and worms. It is the true "lone wolf," solitary in habit except in the breeding season, and an endangered species, thanks to habitat encroachment by man. Several populations are being maintained in zoos, and it was in captive groups that the maned wolves' vocalizations were studied.

At the opposite extreme is the bush dog, which has a low chassis, small rounded ears, and a stumpy but bushy tail. An animal of the forest and forest edge, taking small animals for food, it is highly gregarious, living and hunting in packs. (It, too, was studied in captivity.)

Third is the crab-eating fox, which is actually more like a jackal than what we think of as a fox. It appears to be largely confined to swamp edges where it feeds on crustaceans, the foxes living in pairs but evidently hunting alone.

In each of the three species, infant whines occur in a variety of contexts, all of which elicit care from adults: As they grow older, the infants whine to protest rough treatment. The adult crab-eating fox uses the whine, not surprisingly, as a sign of submission, as does the maned wolf. The bush dog, however, has elaborated the whine into a complex continuum of sounds indicating different levels of arousal. These range from a simple one-syllable short whine to a more aroused collection of whine syllables to a fusion of syllables in an extended whine-scream. The repetitive whine is in fact used by all three species, with displaced pups using it to reestablish contact with parents and vice versa. It is emitted during adult food-sharing and most other situations requiring group contact. Upper-range modifications of the repetitive whine include, in bush dogs, a pulsed vocalization that is louder and has more syllables than a typical repetitive whine, and the siren howl with which the crab-eating fox cuts loose when pair or family members have become widely separated.

Close approach by a hostile or unfamiliar individual elicits a growl in all three species, but only the maned wolf and the crab-eating fox use the bark to signal the approach of a potentially dangerous animal of another species. Unlike those two, which emit single barks, bush dogs bark repetitively when they get angry at one another. Beyond those vocalizations, the maned wolf emits a hum—really a rapid series of clicklike sounds when examined on a spectograph—and a scream, both apparently designed to prevent hostility among themselves when they are at a middle distance; the maned wolf's roar-bark is a long-distance warning, like the roar of a lion.

As has been found in a wide range of mammals, in these three canines an increased rate of repetition of a sound accompanies increased arousal. Similarly, the low, harsh sounds are aggressive, keep-your-distance sounds, while more tonal, higher sounds suggest submissiveness, friendliness: They are what might be called "affiliative" sounds. And, predictably, the gregarious bush dog has the most elaborate set of close-range affiliative vocalizations, matching the need of this social species to express subtleties of mood and maintain contact in the dense groundcover of the forest. The solitary maned wolf, on the other hand, has a more elaborate system of sounds with which to maintain distance between individuals. And lying somewhere in between on the joint continuums of complexity of both society and vocalization is the crab-eating fox.

Q.E.D.

Five

‖|‖|‖|‖|‖|

TRANSACTIONS

In which tales of deception among birds, monkeys, and even the chimpanzees of Gombe raise some significant questions; several evolutionary engines are suggested as the power behind the rise of intellect among animals, including human ones; the reader detours usefully back into a beehive; and a Machiavellian world-view is adduced in biological affairs.

Definitions are often the bugaboo of science and philosophy. It is possible to define animal vocal communication in many ways, and in each case the definition tends either to limit or broaden the way we see the nature of this communication. A wrong word choice can produce an error in emphasis, if not actual confusion. We have noted earlier, for example, how biologists have become prone to using the word "altruism" when what they mean is a kind of nepotism, an entirely different matter. Diction can be crucial, and this is the case in current investigations of "deceit" among animals. There is a story about the vervet monkeys that is either apocryphal or at least anecdotal—meaning that it is not regularly seen as part of the vervet pattern and may have been a single instance that was misinterpreted. The story suggests that vervets may be able to lie by manipulating their predator-specific alarm calls. On the occasion in question, there was a fight going on among two vervets. A close relative of the one who was getting the worst of it, seeing this, proceeded to emit the bark that signified the presence of a leopard. All the vervets, including the two combatants, dropped everything and raced up into the trees. The fight was over and, vervet attention spans being short, was not resumed. Even if this really happened, does it amount to what we call a lie, a deliberate misuse of information?

One can say that it does if the bark used for leopards really means just that and nothing more: "Alert! Leopardlike predator!" But if the bark used when leopards and other ambulatory predators are around actually means simply "head for the trees," then it does not amount to a lie at all, but to a generalized command, useful for a variety of situations, however it was learned. (Perhaps the perceiving vervets have merely learned that a trip to the trees is the best course when that particular signal is heard.) There may be no way for vervets to give voice to the specific thought—"Break it up, you guys"— so they merely use the means at hand, which is to send everyone into the trees, or the bushes, or into a bout of searching the ground: Anything would presumably do to disrupt the fight.

▮▮▮▮▮▮▮
THE ILLUSION OF DECEIT

This, of course, is the rankest of speculation based on an anecdote, but the point is that it behooves us to be extremely careful when discussing deceit and lying and the deliberate manipulation of the conventional content of signals, because this implies a glimmering capacity not only to reflect in some manner on the meaning of a signal but then, in a forward-thinking manner akin to planning, to deliberately misuse the sound for another, perhaps even opposite purpose than that for which it evolved as part of the species's repertoire of signals. This in turn implies a thinking process which we would find recognizable in ourselves and, beyond that, a kind of consciousness, a mental image of oneself and others as actors through time past and future. The idea of consciousness is, to speak plainly, a terrible hassle.

There are what might be called neurophilosophers who say that even human consciousness is an illusion brought about by the chemical and electrical pulsing of the brain and other organs. There are others who would say that a sufficiently complex computer is theoretically capable of a kind of consciousness. Such potentially important nitpicking aside, we generally know what we mean by saying we are conscious: We are aware of ourselves, our thoughts and our actions, and our effects (then, now, and later) on others—all of this being qualified by the phrase "to one degree or another." To impute this sort of awareness to nonhuman animals poses some very difficult problems of proof, however we may feel about the subject either intuitively or ideologically. One way to impute it would be to discover "deliberate" deceit among such creatures.

Deception of an *unconscious* kind is a common feature of nature. The moth which, when alarmed or attacked, suddenly opens its wings to reveal a pair of huge, startling "eyes," is being deceptive, as is the butterfly which has evolved to look very much like another species of butterfly that is extremely distasteful to its attacker. It was evidently a nineteenth-century wildlife painter who happened across another evolutionary tactic of deception called disruptive coloration, a kind of camouflage: He painted some ducks in their natural setting and they were hard to see. If you look at a wood duck in the bird guide, it stands out in natty brilliance, but seen on a stream or at streamside in the mottled sunlight, its outline tends to fall apart into separate, differently colored components that are, overall, as variegated as the background.

One could create a vast catalogue of such deceptions—what might be called anatomical deceit. But deception that involves behavior would seem to be of a higher level. The raising of the hackles—what scientists with precision, if a tin ear, call pilo-erection—is a form of behavioral deceit but clearly it doesn't involve any forethought or deliberateness on the part of the threatening animal. Like many ground-nesting birds, the killdeer (a kind of plover) will lead an intruder away from her exposed eggs or chicks with a display of pathetic injury, wobbling and lurching away from the nest with what appears to be a broken or severely injured wing. That this is relatively effective is suggested by the fact that it has evolved separately as a distraction device in several only minimally related bird groups: That is, it must have arisen several times in independent evolution. Is there anything akin to conscious deceit in this performance? One researcher noted that a female killdeer, when lurching off, fairly consistently changes her angle of departure in tune with the angle by which an intruder approached, so as to put maximum distance between the intruder and her nest, suggesting to the researcher that she had a kind of awareness of what was going on, and an ability to adjust her performance to changing circumstances.

The quest for examples of clear-cut intentional deceit has inexorably led researchers to primates—apes and monkeys specifically, most of which tend to lead highly social and highly vocal lives. They are the most like us, the most closely related to us, after all, and we are the preeminent (if not only) deceivers in nature. It is worth pointing out, however, that it is far easier for us to lie and get away with it if the lie is couched in our spoken or written language. Actors, poker players, a few politicians, magicians, and some others learn to gain sufficient control over their nonverbal communication to get away with falsity in this realm: Most of us find it very difficult to lie with our bodies, our gestures and expressions, especially to someone who knows us well—our family, our immediate circle of friends and neighbors or, in primate parlance, our band. Even so, humans armed with speech are so commonly engaged in deceit and in guarding against it that we may have an inherent predisposition to see it in the behavior of nonhuman animals, especially other primates.

Many people who study primates in the wild have stories about behavior that appears to be deceptive, but because they are stories—not regularly repeated, common events in the otherwise patterned events of primate life—they are rarely reported in the scientific literature. (A reasonable scientific consideration: How many "anecdotes" does it take to make a pattern?) One such observation was made by a Swiss scientist, Hans Kummer, studying a troop of hamadryas baboons in which, as is typical, a single male controlled a harem of females, constantly on the watch lest one of them attempt to approach and interact with any of the other males. In this case, one female was seated within view of the old harem-master and, still seated, inched her way across the ground some six feet—during a period of about twenty minutes—until she ended up with her head and upper body visible to the harem male but her hands concealed behind a rock. Also concealed behind the rock was a young male whom the female groomed.

In a different study of different baboons, a young but strong adolescent had manhandled a younger one, which screamed loudly. As usually happens, a number of adults, including the young one's mother and the male leader, came running up, heading for the adolescent bully. Instead of running off or showing submission, he stood up and looked around at the surrounding hillside, just as one does when one has seen a predator or an alien group of baboons. The adults came to an abrupt stop and looked intently at the horizon too, the result being that the chase ended abruptly. The researchers, too, scanned the horizon and saw nothing.

In the first instance, it would seem that the female had some sort of picture in her mind of what she looked like to the boss—that is, partly concealed—and counted on her concealment to keep him from noticing that she was doing something he would disapprove of strongly. In the second instance, the young adolescent baboon might have had what amounts to a creatively deceptive thought: to distract the adults by assuming the posture that heralds the arrival of a foreign visitor. Another explanation, of course, is that some time in the past the adolescent had had real cause for alarm at the same time that he was bullying a younger baboon and had noticed the effect of his alarm posture. Thus it might have simply become reinforced behavior, as the more mechanistically inclined would put it.

Yet another kind of deception has been spotted among baboons, in this case gelada baboons, which have a similarly tyrannical arrangement between a leading male and a harem. In this instance, a female and a male were out of sight but within earshot of the harem male and proceeded to copulate, which is almost always accompanied by a bout of mutual shrieking. In this case, the baboons withheld the screams, in a sense withholding information which, as anyone knows who has watched a few courtroom dramas on TV, is tantamount to lying.

And there are accounts of outright misuse of vocal signals. Two British researchers, Richard Byrne and Andrew Whiten, observed a juvenile baboon approach a female adult and watch her intently as she dug up a root. There were no other baboons in sight in the surrounding grassland. The juvenile looked around and then screamed loudly. Within seconds, the juvenile's mother streaked onto the scene and chased the adult female out of sight, leaving the juvenile with the root, which he promptly ate. The juvenile was too small to dig up the root by himself and, since this was behavior that he repeated with several different victims in several days, it smacked of what could be called tactical deceit. On the other hand, it is possible that at some point earlier, the juvenile had approached an adult with food and had been threatened. Emitting the scream typical in such situations, the juvenile saw that his mother arrived and drove off the aggressor, leaving the juvenile with a piece of food as a reinforcing reward. Apparent premeditated deceit could actually be mere conditioning of a response to a stimulus. For many of these one-shot accounts there is an alternative, killjoy explanation that is equally plausible, if perhaps equally tortured.

Tales of deception abound in chimpanzee studies. A Dutch primatologist has described how some chimps are masters at emotional censorship, approaching another chimp they intend to thrash without any of the usual gestures or expressions of aggression. A chimp in captivity was trained to locate a piece of food that was hidden in its enclosure during its absence. It would show a friendly trainer the location of the food, aware that the trainer would share the food with it. But it would not only not lead an unfriendly trainer (one who took the food for himself) to the source but instead would lead him to other places, away from the source.

THE RISE OF INTELLECT

If there is deceit—that is, if an animal spends a certain amount of time pulling the wool over the eyes of others—then it is reasonable to expect that animals spend a certain amount of time trying not to be deceived. As a practical matter, counterdeception would be even harder for a human observer to perceive than deception itself. There is one fairly well-known account of counterdeception, arising from Jane Goodall's long-term study of chimpanzees in Tanzania's Gombe National Park. A young male named Figan spotted a banana that researchers had placed in a tree and which other chimps had overlooked. Beneath the tree, a big male named Goliath was resting. Figan glanced briefly up at the banana, then at Goliath, and then moved away to a spot from which he could not see the banana, thus avoiding the problem of having to suppress a desire to glance at the banana again and give its location away to Goliath. In fifteen minutes, Goliath got up and left and Figan immediately went back to the tree and collected the banana. That's deception. But in a similar situation, an adult male was about to eat some fruit when another male appeared nearby. The first male walked away from the food and sat down, looking around as though food was the furthest thing from his mind. The second male walked off out of sight, but then hid behind a tree and watched. As soon as the first male reapproached the food, the second male left his hiding place, shoved the other chimp aside, and took the food. Counterdeception?

While the evidence of deception (and especially counterdeception) remains sparse and still subject to interpretation, it has led to an intriguing theory about the origins of the human intellect—one that would appeal to lawyers certainly, though not to people with a rosier outlook on human nature. Stated briefly, it says that out of attempts to deceive on the one hand and attempts not to be deceived on the other, a kind of cleverness arms race could have taken place among such highly social creatures as primates. The capacity to deceive implies that the deceiver (animal A) can attribute to the dupe (animal B) an awareness or quasi-thought process similar to his own. It also implies that A knows that B will react in a predictable manner to A's signals. It implies that A has a kind of self-awareness and a capacity for fairly high-level prediction. Counterdeception implies even more: that A wants something, that A will mislead B about it, that B knows that A wants that something and is going to try to mislead B about it, and so forth . . . leading to an increasing sophistication in reading the probable intentions and thought processes of others, a game of ruse and counterruse and counter-counterruse with which we are all too familiar in many of our everyday informal as well as formal negotiations.

Other kinds of arms races have been proposed as candidates in the achievement of what we call intellect. One is known as "reciprocal altruism," and the most frequently cited instance is from studies of baboons in the Gombe reserve, wherein one animal gives aid to another that is not closely related, on the principle that the other will one day return the favor. In this case, a male would often be seen enlisting the aid of a second to attack a third who was consorting with a female. Usually the male enlisting the help would wind up with the female while the enlistee wound up in a brawl with the third male. As it turns out, the enlistee usually got the same kind of assistance when he sought it: In fact, the result was long-lasting alliances between males, evidently based on mutual confidence. Natural selection would tend to favor such a relationship if the two-baboon gang makes out better than individuals who do not engage in such practices, which (in terms of sowing oats and possibly other gains like appropriating food resources) they indeed would . . . if the confidence is well-placed.

But what if the enlister of aid from others is a con artist, a cheat, and doesn't reciprocate? He would enjoy the fruits of the system—spreading his genes around—without suffering any of its costs. Genes for selfish behavior, if they exist, would be selected for, while altruistic behavior would be selected against. But this assumes a continuing stupidity on the part of altruists: It is more likely that as selfishness increased, altruism would decrease and assistance would be withheld from the chronic cheater, who might well be ostracized from the group, the penultimate bad news for a social animal. Eventually some sort of equilibrium might be achieved.

Or, theory suggests, the advantages of cheating would still be worth the risk of social ostracism, putting a premium on cunning, which in turn would put a premium on vigilance on the part of the others. The two strategies would drive each other in a continuing arms race heading for the highest form of intelligence. Such a theory proposes a group that is small enough so that all its members can recognize and know one another, that is stable enough so that there is time for them to develop a "book" on each individual, and that it is composed of both near kin and relatively unrelated members of an animal species that already has developed the cognitive capacities necessary for the above: that is to say, primates, notably baboons and chimps among living primates and presumably the apelike predecessors of hominid lines. In this scenario, there is a growing sense of social feeling—trust and understanding as well as deceit and suspicion.

Why wouldn't this same sort of arms-race effect have taken place among many other social animals, elephants and wolves, for example, both of which show a considerable level of social coherence and even cooperation? Note the baby-sitting common among elephants, as well as the long period of childhood and thus learning time, which is considered prerequisite to the development of an "intellectual" creature. Probably one factor is the degree to which an animal's physical repertoire of actions is flexible. An animal whose legs are all used for walking and running (or flying and hopping) has less flexibility to manipulate the resources of the environment than one whose front legs can be put to a variety of tasks like grasping and carrying. (That there are yet other forces at work here besides the flexible use of limbs is suggested by the highly social lives of dolphins. Dolphins are just about as dexterous as horses). In any event, suppose that reciprocal altruism, associated with the practice of having and raising one or a few offspring at a time that require a long period of education by adults, led to intellect. It is a risky "strategy." The altruist could be duped, the investment in a limited number of offspring and in a long education could be easily lost, either by death of the offspring or its teachers. But as one zoologist, Colin Beer of Rutgers, has put it, "nature does not look beyond the immediate odds when placing her bets." At each stage along the way in the arms race(s), the benefits must have outweighed the costs.

At this point the reader may ask what all this has to do with the question at hand: animal vocal communication. And the reply: a promise that its relevance will become clear as matters unfold. A premise here is: *What we think of as intellect has to have a biological function or it wouldn't have come into existence at all.*

Similarly, our form of vocalization—speech—is also unique, but has to have been a biological asset in each step of its evolution as well. Since it is impossible to imagine our intellect and our speech to be unrelated, theories about the rise of intellect will presumably shed light on the nature of animal vocal communication.

An immersion in biology does provide one with a somewhat ornery, if not downright contrary, way of thinking. The evolutionary achievement of intellect is not only a biological matter but also, viewed biologically, a mixed blessing—merely another adaptation to the world. Indeed, from the standpoint of biology, it is altogether reasonable to say that we are too smart by half, that intellect has taken us too far in some ways and not far enough in others. Seen dispassionately as a biological presence on the planet, we are without question a rogue species, with at least a fifty-fifty chance of creating catastrophe that is, in human if not geological terms of time and process, irreversible. The creative side of evolution—the proliferation of life forms—is matched only by its destructive side—extinction, to which no species, not even our own, with our vaunted arrays of intellect-based technique, is immune. It is this technical capacity to challenge and temporarily overcome the ecological constraints of the environment that often draws our attention and warrants our pride. But not only may this talent eventually be our downfall; it may be that our intellect came into being in the service of an entirely different function. It may be that all our technical mastery is merely a byproduct of the biological origin of intellect. The notion of this other function is what underlies the various theories we have been discussing.

■I■I■I■

SOCIETY AND BARGAINING

The notion, which is chiefly that of a psychologist from Cambridge University (England), Nicholas K. Humphrey, begins by pointing out our tendency to look upon creative intelligence as being evidenced by "practical invention," new ways of dealing more effectively with the external physical environment: i.e., man the tool-user. We are greatly taken with the fact that a captive chimpanzee, confronted with a banana suspended above his reach and a couple of boxes lying round his enclosure, will soon figure out how to make a staircase or ladder out of the boxes and reach the banana. We are equally taken (and in some ways taken aback) by the fact that chimpanzees have been found to use tools in nature as well. While foraging, a chimp may pluck off a twig, peel it, fashion it to a proper length and, upon arriving at a termite mound, insert it into a hole in the mound, twiddle it, and withdraw it along with some tasty termites. Unquestionably, such a technological advance is beneficial, and natural selection would, one can guess, favor its development. Nor is such tool use, or any new technique taken up by any species (in hunting or navigating or whatever) to help it master the environment, to be sniffed at. But, Humphrey argued, this doesn't explain much about intelligence and its biological function, for the reason that subsistence may, in some species, call for an accumulation of knowledge (about the whereabouts of food in changing seasons, for example) but not for a great deal of deductive reasoning. Each chimp does not need to reinvent the termite stick but merely to watch an older chimp make and use one. And the evidence suggests that such subsistence techniques as the chimpanzee's termite stick are arrived at in the first place by trial and error—basically a lot of fiddling around, thus explaining the first instance of this behavior, rather than what we would call premeditated invention. If the social

structure of a species is such that there is time and op-
portunity to learn by imitation, then there is little need
for individual creativity.

"Paradoxically," said Humphrey, "I would suggest
that subsistence technology, rather than requiring intelligence, may actually become a substitute for it." He
points to the natural lives of gorillas which, if left alone
by humans, have the simplest life of any creature in the
forest: plenty of food, practically no predators, nothing
much to do but eat, sleep, reproduce, and play. Once
such a *modus vivendi* has been established, through a
knowledge of the environment (and even a difficult environment), the need for creative practical invention is
very small. The millions of years during which hominid
and human hunters and gatherers subsisted in much the
same way with basically the same tools and technological
basis attests to this. So what is all the brain power about?
Chimps and gorillas test surprisingly well on practical
inventiveness, which they rarely if ever apply in their
natural lives to the daily exigencies with which their
physical environment confronts them. They don't need
to. Indeed, how many humans go through a happy and
productive life without ever inventing anything of a practical nature, any new way of doing anything at all? Like
chimp society and gorilla society, our community tends
to provide a medium for the cultural transmission of
much of the knowledge that is needed for us to survive,
along with the protective environment in which we can
learn it. Chimps can imitate; humans can also read in-
struction manuals. The "chief role of creative intellect,"
Humphrey proposed, "is to hold society together."

Whether we take deceit or the less cynical reciprocal altruism as the engine, in a complicated society like that of most primates, each member benefits from both preserving the overall structure of society and outmaneuvering others in that society. Such primates simply have to be calculating. In a game that includes plot and counterplot—however low-key it may be—one has to be able to calculate the consequences of one's own behavior, the likely behavior of others, and the balance of advantage and loss, all in a situation full of moving targets. That is to say, the context in which one calculates the consequences of one's actions is changing all the time, not least as a result of those actions. If A does one thing in some sort of attempt to influence the behavior of B, B may have several options. Depending on which one B chooses, A's options also change, like a game of chess, in fact, or even like a simple child's game such as Go Fish. Thus the sheer necessity for social skill brings into play a kind of forward planning, "a level of intelligence which" Humphrey suggests, "is unparalleled in any other sphere of living."

Why should such societies be complex in the first place? If a society's role is to provide a technical school for its young (a prolonged period during which the next generation can learn the techniques of subsistence from its elders), then natural selection can be imagined to favor a more and more prolonged childhood and also the existence of elder relatives in the society, creating a considerable generational diversity. And with the diversity of age groups comes a diversity of individual needs. The nurturing of young calls for fairly widespread tolerance, among other things, but in times of short food supply, for example, tolerance of the hungry cries of the young may become strained. There is always, in such a society, conflict of interest and politics. And in such a realm, the advantage to an individual (and his progeny) who can outwit the others without totally disrupting the protective fabric of society is obvious. Each such advance would also ratchet up the level of social complexity—at least up to the point where intellectual prowess can go no further. What would stop it?

There may be physiological restraints such as brain size. But there is also the ultimate constraint: No matter how bright one is, one does have to make a living. The film *Dangerous Liaisons* portrayed a group of eighteenth-century courtiers who evidently had nothing whatsoever to do but use all their wits to invent and carry out games of sexual politics. Few of us, and no nonhuman animals, have the luxury of turning social skill into full-time and exclusive activity, but in any social situation a varying amount of time must be spent in social caretaking—which is not directly productive of such basics as food. So, to be biologically advantageous, a social system must provide, via its educative functions, better sustenance techniques to make up for the lost time. As Humphrey says, "if an animal spends all morning in nonproductive socializing, he must be at least twice as efficient a producer in the afternoon." At some point an equilibrium tends to be reached, and the point would differ for each species. It is interesting to note that the chimpanzees of Gombe become much less sociable in years of poor harvest.

It may be that when protomankind moved out into the open savannas, new dividends were to be paid by this environment to a creature already well-ratcheted up the scale. The ability to understand the doings of new animals like lions and gazelles, to know the places where hyenas were likely to scavenge and leave a bit of nutritious bone marrow behind—all this would imply the need for more technical knowledge that in turn implies even longer periods of schooling for the young, implying even greater social complexity.

Up to a certain evolutionary point, it is not difficult to see how animal communication plays out its role in this hypothesis (or scenario, as some like to call it). The Motivational-Structural rules that govern growls and whines predict that in more complex social situations, there will be more gradations of vocalizations along the continuum, both in number and in subtlety. We can look back again (or sideways) to the vervets. They have a relatively sophisticated means of warning one another about the different kinds of danger they can expect from predators. This is a subsistence skill or technique, and it appears to be learned by the young through a certain amount of trial and error and comparison with adult example. But by far the most common vocalizations emitted by vervets are grunts—a sound something like clearing one's throat with the mouth open. All the grunts sound pretty much alike to human ears, but armed with spectrographs and playback tapes, researchers have found a great variety of grunts, different sounds for greeting a dominant monkey or a subservient one, for moving across open territory or for nervously contemplating such an activity, for greeting a member of one's band or an alien group. And each grunt elicits a different response, a different turn of the head, or a different posture or gaze. The grunts, as predicted, appear to represent a continuing series of readings of social context and politics, readings that are as precise and as complex as the vervet intellect has gotten so far.

Assuming that mankind arose out of this sort of situation, it is reasonable that the human intellect is suited primarily to thinking about other people and their institutions. Humphrey suggests that humans tend to view the world as a collection of social problems—that is, matters with which we can transact. As in any social transaction, the object itself keeps changing, partly depending on what we do. The transaction is indeterminate. Thus, for millennia, mankind has *bargained* with nature by prayer, sacrifice, and other forms of persuasion. Our current experiment in objective and logical thinking about the inanimate world—called the scientific revolution—is relatively new, maybe 500 years old in any well-developed sense of scientific reasoning, and, as Humphrey points out, while we believe we ought to think logically and objectively about inanimate or nonhuman matters, we still tend to think in terms of social transactions. Such thinking could have been helpful, in fact, in such things as chemistry. While no amount of persuasion or bargaining is going to change lead to gold, there is a kind of ongoing processual thinking that is creative in such an endeavor. Another example is plant cultivation—hoeing, fertilizing, pruning, and other horticultural activities are related to the plants' ongoing and emergent properties, which are in turn affected by the activities of the horticulturalist. Plants don't seem to respond to human conversation, but there is something in their cultivation akin to a social transaction. "Thus," says Humphrey, "many of mankind's most prized technological discoveries, from agriculture to chemistry, may have had their origin not in the deliberate application of practical intelligence but in the fortunate misapplication of social intelligence."

And if it has been difficult for mankind throughout his history, including his recent history, to be totally objective about a lump of copper and a vial of acid, think how difficult it is to be objective about our fellow creatures— especially those that seem the closest to us, what are called the higher vertebrates, the animals that vocalize. We look on them as both kin and alien, something objective out there, something different, to be understood and at the same time something clearly related to us. In the mirror of their eye, however dark it may be, we see something we can bargain with. And so we try, even in a scientific age, to understand them on our terms. We see their transactions, including their communication amongst themselves, in terms of our own transactions. In their voices we read, or try to read, information. In a sense, we have defined animal communication many times before we have listened well enough to know what we are defining.

▌▌▌▌▌▌

A HIVE OF INFORMATION

As discussed earlier, the idea of information exchange came to animal communication studies largely as a result of the discovery of how bees enlisted fellow members of the hive to forage for nectar. The dance of the bees appears to embody specific information in some way. "Information" had been used in the general sense of "knowledge" until the more mathematical engineering definition came into play—a quantifiable reduction of uncertainty in the face of several options. Both meanings had what scientists call heuristic value—pointing the way to further investigations, new hypotheses. One could track certain kinds of activities in a complex system the way an engineer creates a flow chart of energy through a system, like the wiring diagram for an appliance. Precise relationships and patterns could be put into sharper relief.

But information theory, like a wiring diagram, is only a model. It is not the real thing. It is possible, for example, using extremely sophisticated electronic cartographic methods, to produce a real-time map of the firing of neurons in a brain and relate this to thoughts and come up with the notion that we think electrically. It is also possible to map the flow of a host of hormones, peptides, and other proteins in and out of the brain and relate these flows and arrivals to thoughts, coming to the conclusion that we think chemically, and with what amounts to our entire endocrine system and not just the brain. Both are excellent models of at least part of what

▐▌▐

is going on, both have heuristic value, and both will almost surely be superseded by a more complex model that fits all the phenomena of thought more closely. But the model will not ever *be* a working brain, or brain/body continuum. In the same way, information is a very slippery item, perhaps even slipperier. Unlike electricity, it does not have a charge; unlike a protein, it does not have mass and shape. It is not a wave. It can perhaps be thought of as a property of objects or events, inherent but inert, given birth to only in the course of some process or transaction. A strand of ribonucleic acid can be said to contain the genetic information needed to produce, say, a cranberry bush, but if it were to exist in a medium in which there was no other material around to latch onto the individual ports along the strand and become cranberry-bush proteins and cells, then one might question whether there was such a *thing* as information in the RNA strand as an independent entity.

That much-mentioned tree falling in the forest-that-has-no-ears nonetheless emits something physical: sound waves, which exist independent of ears. On the other hand, a book lying eternally unread on a table emits nothing that could be called information. These philosophical matters are raised, however naively, to raise yet again the point that "information" can get out of hand as an explanation of what is going on in animal communication. Information can balloon not just into an independent something extant in a situation—an interaction between two birds, say—but, if not kept very carefully strapped down in our minds, it can come to be thought of as very nearly the *reason* why the two birds are there in the first place. It can become reified, and taken as independent player on the stage.

For example, the colony of individual bees in a beehive goes about its multifarious business as if it were one organism, and such situations have been called superorganisms. Each bee operates independently within the system, though no bee could survive alone. Most individual bee activities benefit the entire hive and therefore the queen, to whom all the other bees are genetically related, but for the most part beehive activities are decentralized. There is no tsarina gathering information centrally and passing out orders, the way a brain tends to do. Bees spend about 30 percent of their time wandering around the hive, tending to what needs to be tended. They also spend a good deal of the time in physical contact, pulling and tugging at each other, evidently communicating one thing or another. A recent article by Cornell's Thomas D. Seeley pointed out that bees communicate "indirectly, through some component of the shared environment." For example, several bees may be involved in constructing a given cell in a beeswax comb, yet they "never need to come together and exchange information directly." Instead, all the information needed is embodied in the cell wall under construction. One bee deposits a bit of beeswax and goes off; another comes along and does a bit of appropriate sculpting. The growing wall, in this sense, becomes a channel for the transmission of information, and the bees (without contact) are thus engaged in indirect information exchange along an information transmission channel (the wall).

Similarly, if a hive gets too hot, bees begin fanning

their wings to pull cooler air into the area; if it grows too cold, the bees produce heat by contracting their flight muscles. The temperature of the air, in this case, is seen as a communication channel in regard to the colony's heating and cooling needs. Again, when a foraging bee returns to the hive, its nutritious freight is off-loaded by another worker, which takes it to the honeycombs to be deposited. Usually a food-storer bee arrives to do the off-loading within fifteen seconds of the forager's arrival: "then she [the forager] knows that there is little nectar coming into the hive and little honey stored in the hive," so she hustles off to dance and recruit other foragers. But if the storage-bee is delayed—up to 100 seconds— before arriving to unload a forager, then the forager knows that the hive is getting full, and gives up recruiting others to get more nectar. (The delay means the storer bee has had to look high and low for storage space.) Thus does the forager receive precise information about the fullness of the hive's stores at any given time. Seeley says that such transmission of information is accomplished by a "cue"—an action that carries information only incidentally—as opposed to a "signal," such as the bee's dance, which has obviously been shaped over time by natural selection to carry precise and specific information and has no other purpose.

It is instructive, of course, to discover the connection between unloading time and continued foraging activity—a connection that might easily have otherwise been missed. And it may be helpful in a metaphorical way to think that a forager, impatiently standing in line, realizes that the hive is full of honey because it is taking forever for old whatsername to get back here, but it really doesn't seem likely that there is anything like "information" about the actual state of the honeycomb, however incidentally encoded in the storage bee's activity, that is "transmitted" to the forager. The *ultimate* effect of forager/storage bee timing may be that the foragers adjust their activities *as if* they knew how full the hive was, and this would seem to embody some (slightly paranormal) substance or property called information. But each individual forager/storage bee interaction is almost surely some other matter—at the *proximate* level, that is, what goes on in that moment when a forager bumps into a storage bee on the loading dock. They are not gossiping in Bee about the state of the warehouse.

Similarly when a whole bunch of bees starts to fan their wings, a researcher knows that the temperature of the hive has probably gotten up over 36° C., but the bees almost surely don't know that: More likely, they are simply hot, and wing-fanning is what hot bees do. For a patrolling bee to put in its two cents' worth on an unfinished cell wall barely seems to qualify as the encoding of information by one bee into a cell wall, information that is thus indirectly transmitted to and then decoded by another bee. Bees probably tend simply to put in their two cents' worth on unfinished walls, the overall, ultimate effect being a honeycomb. This is not to say that adding up all the potential pathways for information in a beehive doesn't lead to certain valuable insights, even if the pathways are incidental and the information "rudimentary," such as that which is incidentally "encoded" in cues or even in the environment. It is possible and, up to a point, heuristic to see a beehive as a web of information channels designed over eons by natural selection. But in the process, it is also possible to lose sight of the fact that what the hive is really full of is bees.

A difficulty in thinking about the "meaning" or the "suggested response under the circumstances" that is inherent in events or objects, and giving it the name "information," is that we begin to think of it in terms of the way *we* typically use information. And this is especially true in considerations of animal vocalizations. Suppose a man walks up to you and says, "I want you to jump off this wharf into the water. If you do, you'll find a treasure chest down there." That is what we would call information (of various sorts). On the other hand, if a man walks up to you and pushes you into the lake, we don't really need to consider that information, even though it may lead to the same ultimate result.

MANIPULATORS AND MIND READERS

In any event, struck by some of the difficulties involved in attributing "information"—especially semantic information, wordlike meaning—to the signals of animals, two British scientists, John Krebs and Richard Dawkins, looked for another model of what animals are actually doing in the course of communicating with one another. They came up with the notion of manipulation and mind reading as the keys to understanding the transactions involved in animal communication.

Even the simplest organism manipulates its environment—for example, making energy for itself by converting part of its environment into nutrition. On a more
sophisticated level, an otter will collect a rock; then,
floating on its back, it will pound the shellfish it is holding on its stomach with the rock to open its shell. In this
sense, the otter is manipulating or exploiting the very
lawfulness of the environment: Rocks are harder than
shellfish shells. Similarly, an animal exploits for its own
purposes that extensive part of its environment made up
of other animals of its own or other kinds. The male
otter, for example, can be construed as using a female
otter to bear and raise his offspring. There may be an
initial reluctance on the part of the female to mate with
this particular male, or in fact any male, but the male who
is successful exploits the lawfulness of female otter behavior, cozening her and coaxing her by means of a
series of behaviors and signals, including vocal ones, to
put aside her sales resistance and get on with what he
wants her to do. Of course, it is ultimately what *she* wants
in the sense of natural selective pressures working on her
and her genetic potential. But at the proximate level, she
needs to be sold. The female behaves in certain "lawful"
ways, as well as a rock does. In employing certain kinds
of signals (here meant not so much as signs or gestures
communicating information but as anything that acts as
an incitement to action), the male exploits the expect-
able lawful behavior of the female and puts her muscle
power to work in his behalf. In this sense, there are
lawful means at the male's disposal as well as unlawful
ones. If he were simply to employ main force, it might

work, or the female might flee, or she might at least remember this particular brute and avoid him thereafter. (*Think about all this in the light of our concept of the importance of the perceiver—the audience determining the comic's routine. It is the perceiver—here the female otter—that sets the parameters of lawful behavior.*)

For the male to exploit, to manipulate, his victim successfully implies another role: mind reading. He must be able to read her intentions as they change in accordance with his manipulative efforts. Similarly, for the manipulation to succeed, the female, too, must be able to read the male's mind, understand his intentions. (Krebs and Dawkins point out that the term "mind reading" is used very loosely here, leaving aside the exasperating question of whether *any* animal has what we think of as a subjective, self-aware, self-conscious mind like ours. Here "mind" is used in the sense we mean when we ask, "What's on your mind? What do you want to do?" In transactions between nonhuman animals, presumably this kind of mind reading is possible only through the registering of the other animal's signals. That dog, for example, has pulled his lips back from his teeth: Therefore he might bite.)

A biologist's mind needs to be a bit Byzantine, and it is worth taking off on what seems like a digression here, arising from the threatening dog with its lips pulled back. For such a signal to mean anything, it has to have a certain predictability: It, or something very much like it, had to have been in use over a long period. It is easy to see how such a simple gesture as pulling back the lips came to have meaning. When a dog bites, it pulls its lips out of the way of its teeth by necessity (lest it bite its own lips), so the two acts are clearly associated. Simply pull-ing one's lips back suggests the imminence of a bite. Similarly, a crab may raise its claws in order to position them to pinch a predator. When it is molting and its claws are useless, the mere act of raising them is still taken as predictive of attack, as a threat, and may ward off a predator at least momentarily. The crouching be-havior of many female songbirds in courtship, too, is probably derived from the food-begging crouch of juve-niles. The point is that a signal, including a vocal one, has to affect the nervous system of another animal in a certain way for it to be effective. It is not difficult to imagine a bird, accustomed to reacting to its offspring's crouching behavior (which is taken to be a sign of ap-peasement), reacting in a not altogether different emo-tional way to crouching by an adult female. It is harder to imagine a brand new signal—a mutant—having such an effect. It is possible, for example, in our fast-changing world, to imagine that sometime after the turn of the century, convention might dictate that a man, upon re-ceiving a flirtatious look from a woman, must race to the coat closet, put on three overcoats and tap-dance on a table, but if one were to do this today, one's intentions could rather easily be misinterpreted and the flirtation end. There are in fact mutant mice that will suddenly break into a mad waltz, rather like a series of random twitches, and these could conceivably one day become meaningful signals, but for now they seem to mean very little to other mice. Moreover, even in a highly evolved signal—one that has accrued symbolic meaning—the original function may still be present. For example, a gland that now emits a sexually attractive chemical could have evolved from a sweat gland, and while the chemical is different and serves a new purpose, its excretion might still serve as a cooling device too. On the other hand, the original function may have been totally lost: Certain ducks in courtship point in exaggerated ways at their feathers with their bills, no longer actually preening them, or even touching them. As with the apparent vomiting of ravens, this process is called ritualization.

In sum, a signal should demonstrate one's state of mind, one's intentions, so that the game of manipulation/mind reading can continue. It goes without saying that in many cases a single individual is both manipulator and victim, playing two roles alternately if not virtually simultaneously. One's signals need to fit convention to be understood as such. But there are situations—poker games, for example—where showing all one's cards up front is a significant disadvantage: Bluff is often called for. But there is also a problem with bluffing: If you do it all the time, it tends to be taken less seriously. Researchers have found, in studies of various birds (blue tits, grosbeaks, and skuas, the latter being large maritime raiders), that there is a very low correlation between many of their threats, both vocal and postural, and the onset of an attack. Not only are threats often inaccurate predictors of attack, but perceivers of the threats often fail to retreat. Indeed, among blue tits, a threat is often followed by the retreat of the *threatener*.

These findings, along with similar studies among certain fish, suggest several things. One is that a game theory approach may be right: It is not to the animal's advantage to signal its *longterm* intentions at the outset. Instead, the threat display may be a way of indicating "I might well attack you, but I might not. I want my uncertainty to make you uncertain." (But then, following this logic through yet more labyrinthine corridors, if such signals are supposed to indicate indecision, why do so many of them become ritualized, exaggerated, stereotyped?) Another insight arising from these studies is that an animal may exhibit a variety of threat signals, any one of which it can use given the circumstances of the moment. In the blue-tit study, it was found that a threat display with a high probability of being followed by an attack during one time of the year had a low probability at another. Quite possibly, for each separate threatening signal in an animal's repertoire, there is some typical frequency with which it is followed by an attack. If the animal uses the signal more often than that frequency, its value as a predictor of attack diminishes and others (the perceivers) pay it less heed. So its use would lessen, while another threat signal would come to be used more often, leading to a series of oscillations in threat choice over time. Thus we have a situation that simultaneously calls for a certain amount of momentary bluffing, for a certain amount of withholding (not of information but of total revelation) *and,* at least over time, for reliability, a kind of overall statistical predictability.

As for manipulation, Krebs and Dawkins point out that this is not simply theoretically possible: It is done often. Researchers manipulate animals by putting what amount to cartoons in front of them, simple versions of the stimuli they see in nature, such as a crude replica of a rival's head on a stick, and get appropriate reactions. Stickleback fish grow fiercely hostile at the sight of another male red stickleback, but they will also get furious when a red truck goes by the window. Krebs and Dawkins suggest that we may think sticklebacks are pretty stupid, getting upset by a distant truck simply because it is red. Do the fish really think the truck is a fellow fish? But, the researchers point out, a man can be sexually aroused by a picture of a naked woman. The man is not fooled into thinking he has the real thing; the picture merely contains enough *in common* with the real thing to affect his physiology via his nervous system. In such ways, many animal signals, including vocalizations, can be seen as vehicles for manipulation. Mind reading, of course, would have coevolved alongside manipulation, another arms race. (Humphrey has suggested that the constant attempt to read ever more subtle cues as to another animal's intentions, as revealed by signals and inadvertent clues—like increased breathing rates and so forth that could hardly be classed as evolved signals, but are nevertheless interpretable—could have led to the whole faculty of subjective human consciousness and self-awareness, the better to read the minds of others.)

What does an animal do if its mind is being read? It depends. One might respond in kind to a threat. But it may be to one's advantage to be mind read. A female entering the territory of a male might be roundly attacked, depending on whether the male is in an aggressive or sexually excited mood. Quite obviously it is to the female's advantage if she can read the male's mind and tell the difference. But it is also to the male's advantage to have his mind read: On the one hand, he avoids a costly fight, and on the other he gets to copulate. The male is a willing victim of mind reading. The cat stalking a bird may be a willing victim when the bird, having read its mind, emits an alarm cry, thus letting both parties know there is no further use to the cat's stalking. Out of the coevolution of manipulation and mind reading, signals would evolve. Taking it a few steps further, Krebs and Dawkins suggest that in situations where there is resistance—sales resistance, as they call it—signals would tend to become more like the essentials of our advertising. That is, they would become repetitive, stereotyped, more noticeable through greater amplitude (noisier, in other words) or exaggeration, and symbolic. (They note that our advertising messages tend to display these qualities and not, to any great degree, information.) On the other hand, where there is cooperation rather than sales resistance, the signals tend to be muted, quieter, less exaggerated— the call notes between mated birds, the contented purring of cats and other carnivores among their kin, the sounds and gestures of sociability. Krebs and Dawkins compare them to the conspiratorial whispers of a couple deciding it is time to leave a dinner party.

One of the appealing parts of the Dawkins-Krebs concept of manipulation/mind reading is that it puts a new emphasis on the *transactions* between communicating animals. From the standpoint of information transfer, there is a tendency to think of the sender of the communication as the active participant and the receiver as largely passive. Animal A encodes its meaning into some sort of signal and shoots this over to B, who is sitting there passively like a computer waiting idly for a command. In comes the command from A, B's brain hums a while, decodes the command, and does what it is told. To imagine, however, B actively wondering what A is up to, and maybe even pulling one or two on A at the same time, isn't just more interesting; it also is more attentive to what animals appear to be doing.

Six

CALLING LONG DISTANCE

In which the reader goes to a dinner party and then to the tropics for a fresh veiw of life; spends a considerable time with the birds, albeit with a brief sojourn among frogs; learns of competition, energy conservation, and (wistfully) romance; and has a glimpse of the relentless perseverance of the scientific mentality, once it is awakened.

A long time ago, by way of making a point, the physicist Edwin Schrodinger put a cat in a box along with some poison that could be released by the emission of a single radioactive particle and, given the hypothesized unpredictability of individual radioactive particles, claimed that there was no way in the world to tell if the cat were dead without opening the box. He went on, with the daffy reasoning of quantum mechanics, to explain that, in fact, the cat would be both dead and alive—until the moment when someone opened the box and one state of being somehow gained instantaneous hegemony. That, of course, was a mind game, the sort physicists like to play (and possibly one reason why there are so few physicists). But anyone can play mind games.

In this one, two friends—in fact, John Krebs and Richard Dawkins—are invited to a dinner party, a very fancy one with more than fifty guests at one long table, lavishly decorated with flowers, candelabra, and so forth. As ten o'clock approaches, Krebs realizes that it is time for the two men to return to their lab to finish up a scientific paper. Under normal circumstances, Krebs would turn to Dawkins and with a twitch of the eyebrow and a glance at his watch, or a low whisper, let his partner know they should leave. But in order to spread the conversational wealth around among her guests, the hostess has seated Krebs at the opposite end of the table from Dawkins. At such a distance, and thanks to the intervening flowers and guests, Krebs cannot catch Dawkins's eye. At that moment, the butler leans over Krebs and says he has just received word that Krebs and Dawkins's lab is on fire. Urgently needing Dawkins's immediate attention, not to mention his cooperation, Krebs emits a shriek something like this: "DAWKINS! WE'RE LEAVING!" This signal is detectable over the buzz of conversation, across the distance, and through the intervening foliage.

The point, of course, is that one of the most important considerations in animal signaling is the distance over which it must travel and the medium through which it must pass. To be sure, loudness does suggest intensity and is often linked to persuasiveness. But there are forms of quiet persuasion: A soft, subtle grunt of superiority can be enough to persuade a young monkey that he is outclassed. It has also been discovered that chimpanzees that find themselves at a great distance from one another in the forest will emit—one and then the other in an alternating series—loud calls known as pant-hoots. These calls communicate the chimpanzees' whereabouts to each other. This phenomenon has only recently come under study so no one really knows. Presumably, notifying members of the group of one's whereabouts is an instance of cooperative communication in the Krebs and Dawkins hypothesis but, just as at their dinner party, the nature of environmental conditions would seem to hold both ultimate and proximate sway over the kind of signal needed, and used.

In other words, plain acoustics.

Krebs and Dawkins suggest that cooperative signals, such as conspiratorial whispers, would reach a certain optimum between the minimum energy expenditure in emitting the signal and its detectability. Too quiet and it can't be heard even from nearby. On the other hand, they say, signals designed for persuasion would become louder, more exaggerated, more stereotyped. In this view, signals that have to overcome sales resistance are seen as not only greater in amplitude but more stereotyped, less subtle, with fewer hard-to-detect gradations of (for lack of a better word) meaning, and thus less subject to misinterpretation. Conversely, in more intimate situations, subtlety is less likely to be misinterpreted.

Perhaps, instead of thinking of muted "whispers" and persuasive vocalizations as the results of two different co-evolutionary arms races, it may be more useful to

think simply of short-distance and long-distance communications. How many nuances can you achieve just in the
tone of your voice if you are cuddled up with a loved one,
explaining the depths and variety of your affection? But
if you are trying to reaffirm your affection across the
teeming lobby of a subway station, you would have fewer
options, and most nuances would be out of the question.
Mouthing each word in an exaggerated fashion, with
unnatural pauses between each word, you would probably shout at the top of your lungs: "I . . . LOVE . . .
YOU! . . ." And what could be more cooperative?

This may all seem obvious enough, but while animal
vocal signals have been recorded, spectrographed, sped
up, slowed down, played back, tampered with, and
analyzed for content and effect, and even had some of
their acoustical properties assayed (for example, the
hard-to-locate, high-pitched, and soft-edged *seet*), the
fundamental nature of these sounds—their acoustical
structure and their acoustical environment—was only
lately taken fully into account in studies of animal communication, even of that most studied of long-distance
animal sounds, bird song. But once the structure of the
captive Carolina wren's *calls* had led to the Motivational/
Structural rules described in Chapter Four, it also
seemed reasonable to look at the structure and acoustics
of bird *song* in an evolutionary context. Perhaps another
direct look at form would settle the question of function
once and for all, and provide a clue to the evolutionary
path that had led to so dazzling an array of melody and
variation on themes.

THE SIREN SONGS OF THE TROPICS

We return to the young biologist pent up in his office in Maryland, yearning to study in the tropics. And indeed such opportunities arose. Being in the tropics provides one with a different point of view on avian affairs than one gets in the Temperate Zone, where most bird song is limited to the breeding season, after which many singers disappear. We dwellers in northerly latitudes tend to think of warblers and swallows and a variety of other migrant songbirds as "ours," but most of them spend more time en route to and in the tropics than with us. It is only relatively recently that tropical biology has become systematic, pointing out an oversight in our studies of such things as birds. Most of the ground-breaking work in the science of ornithology and the insights into such areas as ecology and population dynamics that have resulted had come from European and North American scientists studying European and North American birds, including those that spend only part of the year there. But birds that breed in the north and then go south for winter are not merely tourists in the tropics, idling away their time in some less-than-"real" ecosystem. They simply have two usually quite distinct ecosystems to live in—three, if their migratory route is taken into consideration: two home ranges connected by an avenue.

In the tropics, bird song tends to be fairly constant throughout the year, unlike in the temperate north. Furthermore, singing by females is quite common in the tropics, while it is a rarity in the temperate zones. Carolina wrens, we have noted earlier, sing throughout the year in their northern grounds, and female mockingbirds do sing. But these tropiclike birds are among the exceptions.

The mockingbird also draws attention again to the fact that some singing birds learn new songs or variations of songs throughout life, while others learn for a certain period of time, perhaps a year or so, after which point their repertoire is fixed. Still others have a repertoire that may be severely limited and is totally programmed into their genes, functioning without any detectable learning process. How does one account for this variety in repertoires—both from the standpoint of proximate tactics (the immediate function of singing) and ultimate strategy (what it accomplishes from the evolutionary standpoint of permitting a species to flourish within its particular environmental constraints)? The biological premise that form and function are inextricably linked leads again to acoustics . . . and back to the Carolina wren. All told, there are nine species of wrens commonly seen in North America, but the Carolina wren's nearest relatives, the other members of its genus, are confined to the tropics of Central and South America, where both males and females sing year-round. But . . . acoustics first.

▌▐▐▌▐▐▌

THE DIFFICULTIES OF SOUND WAVES

Any sound is faced with a series of obstacles to its continuation. In the first place, we know that, once emitted, a sound immediately begins to spread like the ripples from a stone thrown into a pond and, in the course of spreading, it attenuates at a known rate. It simply peters out. In a medium free of any acoustic obstacles, sound would attenuate at a rate of six decibels for each doubling of the distance it travels. But most sounds occur in a real world of acoustical obstacles, including the air, the ground, and vegetation. Even the temperature gradients in the air deflect sound, as does air turbulence and wind. Also, any sound has a certain range of properties, such as frequency (pitch), tonality, or amplitude (loudness), and each of these properties is affected differently by different obstacles or properties of the medium through which it must travel.

A tilled field full of soft, furrowed earth has a profoundly different effect on sound traveling over it than, say, an asphalt parking lot. In this it resembles the way sound travels in a room with or without a carpet. A sound moving over breezy grassland will be affected differently than one traveling through a forest densely populated by tree trunks, a place where the air is still and uniform in temperature, thanks to a dense canopy overhead.

If you were an engineer trying to design a sound that was ideal, you would try to isolate each property of the environment and measure its effect on the nature of sounds. This can, to a degree, be done by using what are called anechoic chambers—rooms that are free from echoes and reverberations, and in which every surface contains rows upon rows of cone-shaped projections designed to absorb sound rather than reflect it. The cones mask all the obstacles to the propagation of sound except for those particular obstacles that you deliberately put in the room. In such a place you can manipulate all the properties of sound and, by isolating them and then combining them in a carefully designed manner, arrive at a general prediction of what any sound will have become when it is heard at some particular distance after experiencing a particular variety of obstacles. In other words, sound degrades—and predictably, if you know enough about the acoustical circumstances. An engineer who knows the parameters affecting sound and its degradation can design a sound that will function the way he *wants* given the obstacles it is to face. However, the engineer of nature—natural selection—does not have anechoic chambers. Vocal and other signals have been designed over a long history of natural trial and error, not theory, but it is reasonable to presume that they have gone through some such design process in order to achieve a match of acoustical form and biological function in particular acoustic environments. Again, here we have in mind the evolutionary play in the ecological theater, a context in which to contemplate the acoustical world of a forest-dwelling bird.

The nature of the forest has a great deal to do with how sound works in it. In untouched stands of spruce, pine, and cedar, sounds in the range of 3,000 to 4,000 kHz attenuate little—sounds in this range are hardly disturbed by the trees at all. On the other hand, in deciduous forests it is sounds in the lower range of 1,000 to 2,000 kHz that are least disturbed. It stands to reason that sounds in the higher frequencies (3,000 kHz and above) will bump into such obstacles as leaves and trunks, and the sound become reduced. But some leaves and tree trunks, depending on *their* size, can act as *amplifiers* of sounds in the midfrequencies (300 to 3,000 kHz). What causes the amplification at lower frequencies probably has to do with resonance, but remains a mystery. In any event, a given habitat has its own acoustical rules for propagating sounds most effectively. In the course of evolutionary time, any species will have to have taken those "rules" into account in the course of "engineering" a vocal signal for long-distance communication: The local obstacles and their exact effects on sounds will have been taken into account by both signaler and perceiver.

If what seems an obvious point is belabored here, it is because its heuristic effects are less obvious. For example, in a closed habitat like a well-canopied forest, the obstacles that are present suggest that sounds should belong to a narrow band of frequencies to achieve an optimum propagation—that is, they should be tonal. On the other hand, in open habitats, lower sound frequencies and a buzzier sound would be called for by the "rules" of acoustics. This is often found to be the case, but not always. Anatomy has its own rules, too. One finds mixed correlations between the frequencies birds use in open habitats and the suggested acoustical preference for low-frequency sounds. That is because the ability to produce both low and loud sounds is a function of body size and a lot of birds are simply too small to make the low long-distance sounds that would take advantage of the acoustical optimum.

COMPETING FOR AUDITORY "SPACE"

Another class of obstacles for sound is other sound—the noise of nature in general. An animal's vocalizations must be audible in spite of an array of environmental sound. The rushing of a stream will mask certain kinds of similar sounds, for example, including the sounds of other species of animals as well as those of one's own species (or, as biologists call them, one's *conspecifics*). In this latter instance, an animal must strike a balance between having its calls distinguish themselves one way or another from the calls of its conspecifics and being recognizable to them as a member of their group. Amid the overall noise of the environment, it has first to be able to get across two things: "I am of the species robin" and "I am *me.*" There appears, therefore, to be what could be called competition for "acoustic space," and this is not only an avian problem.

Studies of tropical frogs have shown that one species will simply alter the nature of its calls when it encounters interference from another species whose calls overlap in frequency. The frogs may reduce either their calling rate or their proportion of multinote calls. Yet another species, confronted by the same problem, may increase its call rate and add clicking calls. A Puerto Rican tree frog, the coqui, is named for its call. The first note, *co,* is an unmodulated sound (it doesn't vary in frequency) and has been shown to function in male-male competition for space from which to call in females. The second note, *qui,* sweeps upward in frequency and serves to attract females. Now the frequency of the macho *co* is relative to the size of the frog, and if a male hears another *co* that is close in frequency (within 200 Hz) to his own, he immediately drops the *qui* and deals only with the approaching threat to his local hegemony. So you get a lot of *co*ing until territorial matters are settled. On the other hand, if the approaching *co* is beyond 200 Hz from his, he simply doesn't hear it. His ear membrane is such that it will only vibrate within the narrow range of frequencies emitted by a conspecific that actually poses real competition.

As with the frogs that change the timing of their calls when faced with temporary interference, there is evidence that birds also do this. The least flycatcher and the red-eyed vireo have songs that overlap considerably in frequency, but the vireo's song is long while the flycatcher's song is short. When the two are in competition for acoustical space, the flycatcher won't start a song while the vireo is singing, probably because the flycatcher's short song would be completely masked by the longer song of the vireo . . . but not vice versa: The vireo apparently sings whenever it pleases. Examples abound. White-throated sparrows avoid temporal overlap with others of their own kind, especially near neighbors. A group of skylarks, after long periods of silence, will suddenly erupt into song while flying, but they sing sequentially, not all at once. This suggests that a bird with a comparatively long, drawn-out song might be able to keep others of its *own* species from singing and thus gain advantage in the competition for acoustic space. Indeed, matters could be even more intense than that.

Experiments with Carolina wrens—in this case, seven males kept in separate cages within hearing range but out of sight of one another—show that singing by one male can have a psychologically inhibiting effect on other males. In one study, a caged male began singing and all the rest remained silent *for several days.* After the number one male was removed, a second male eventually began singing, and again none of the others would sing until long after this second bird was removed. So it went all down the line. Since all the birds eventually did sing, it is clear that one bird can directly inhibit the singing of its perceivers.

To return, however, to the complexities of the natural environment. Do birds (as far as they are capable) show a tendency to follow acoustical rules about the "best-suggested-sound-in-the-circumstances?" Evidently yes. Just as a bird may vary its call in one way or another when presented with acoustical competition from another species's song, so birds of the same species vary the nature of their song depending on environment and even on the time of day. As noted, in a forest the air is apt to be relatively stable and with uniform temperature, while an open, grassy habitat may have more turbulent air. In the forest, a tonal song is "recommended"; in the meadow, a lower, buzzy song. There is a ground finch species in

Panama that is found in both mountain meadows and lowland savanna. The mountain population sings with a tonal song but only in the still air of dawn before the sun raises the air temperature, causing turbulence. On the other hand, the savanna birds use a buzzy song throughout the day. Thus do different populations of the same species follow the acoustical rules of their habitats.

Perhaps the most telling example comes from the original case of what biologists call adaptive radiation—the finches of the Galápagos Islands that are now called Darwin's finches because they caught his eye during his trip on the *Beagle* and intensified his belief that evolution was powered by the force called natural selection. The Galápagos Islands were at some point colonized by at least a pair of mainland finches, and over time these evolved and radiated into a variety of niches: From the original ancestral population on the islands, there arose several species, each of which became adapted to a slightly different life-style, based chiefly on food. One finch species took up the food preferences of warblers, and not surprisingly its bill evolved into the form typical of a warbler's bill. Another species took up the food preferences of mainland woodpeckers and, again, its bill evolved into a strong tool for hammering wood. It goes without saying that preference for one kind of food over another also leads to the choice of a particular habitat (where the preferred food is) and, as we've seen, each habitat has its own acoustical properties. It turns out that the "warbler" finch on the Galápagos Islands not only developed a warblerlike bill, but also the sound of certain American warblers. The "blackbird" finch, the "titmouse" finch, the "towhee" finch and the "woodpecker" finch all sound much like their mainland counterparts. This clearly, even beautifully, illustrates the close and sophisticated link between such widely diverse aspects of a bird's life as food niche, habitat acoustics, even social system, and the evolution of signal *structure*.

So in evolutionary terms, birds adjust their songs to their environment and its acoustical properties . . . but to achieve what? Certainly, since the sounds are long-distance communications, distance seems to be one goal. Loudness certainly is one way to achieve distance, but it isn't the entire answer if the caller is trying to indicate how close or how far away it is from a listener. The *perceived* distance would be slightly different if the singer merely turned its head away from the perceiver's direction. Of course, a bird can vary the loudness with which it sings, too. There is also the constraint of size and other matters of physiology. Monkeys, for example, might be able to achieve greater loudness if they could call out in lower tones than they do, but they can't hold enough air to increase loudness at low frequencies, even though some have special resonators in their throats. They achieve the best sound propagation they can given their bodies and the nature of the medium. Similarly, birds apparently sing in a manner to achieve not necessarily maximum loudness but, instead, maximum delivery of sound over distance *with a minimum of degradation* from the acoustical obstacles of the immediate environment.

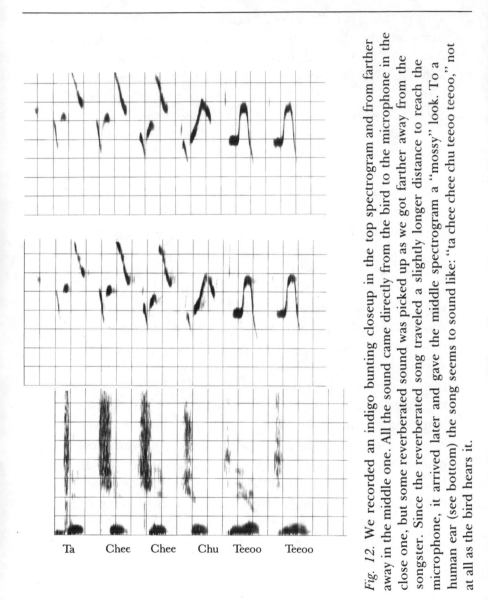

Ta   Chee   Chee   Chu   Teeoo   Teeoo

Fig. 12. We recorded an indigo bunting closeup in the top spectrogram and from farther away in the middle one. All the sound came directly from the bird to the microphone in the close one, but some reverberated sound was picked up as we got farther away from the songster. Since the reverberated song traveled a slightly longer distance to reach the microphone, it arrived later and gave the middle spectrogram a "mossy" look. To a human ear (see bottom) the song seems to sound like: "ta chee chee chu teeoo teeoo," not at all as the bird hears it.

▮▮▮▮▮▮

MAKING THE BEST OF DEGRADATION

Bird song, like any sound, must change (degrade) as it travels, and the importance of the amount of degradation has been shown experimentally with Carolina wrens. A colleague, Douglas Richards, showed that among these birds the amount of degradation in the song as it is perceived, and not its loudness, is what is heeded. The experiment was based on the fact that a Carolina wren has two basic responses to a conspecific's song.

(1) In its most aggressive response it drops everything (ceasing foraging, for example) and attacks the source, be it a conspecific or a loudspeaker. It will search actively for this intruder and won't resume singing until several minutes of silence have elapsed.

(2) In its less intense response, it sings back immediately and stays put, returning to the chore of foraging almost immediately after the other bird ceases to sing.

What is it that inspires the more aggressive response instead of the laid-back one? Richards recorded the song of a Carolina wren at various distances in the woods (mindful that any increase in distance automatically increases the amount of acoustic degradation). Then he played these variously degraded songs back to the bird. Loudness made no difference but, if recorded song seemed highly degraded, the wren chose response (2), singing back and going on about its business. It acted as if there were little to get upset about—in fact, as if the intruder was somewhere outside his territory, worthy of

▮█▮

attention but no cause for making a fuss. But upon hearing the *undegraded* version at the same loudness and from the same distance, the bird attacked. The conclusion is inescapable: The bird used the amount of a song's degradation to get a notion of the distance to its source, and responded accordingly. Subsequent experiments showed that even a degraded song that was *louder* than the undegraded song had the same effect: The perceiver was gauging (or ranging) distance not by loudness but by the degree the song had degraded.

What this might say about such recent communicatory phenomena as rock concerts is anybody's guess, but it immediately raises some theoretical questions harking back to the matter of the information perspective, as well as shedding some light on the true role of bird song. It highlights the difference between the two roles that a bird plays in this situation: the singer and the perceiver. If the perceiver uses the amount of degradation to range a competitor's distance, does the singer produce songs that degrade "predictably," thus cooperatively providing accurate distance cues to listeners and conveying the honest, uncertainty-reducing information as one might expect from the information perspective, especially with both birds benefiting from the exchange by using up the least energy in a territorial squabble?

Or (as the logic of natural selection suggests) is a singing bird out just for itself, providing accurate distance information only when it provides the singer with a net gain of some kind over its competitors? In the latter case, singers should typically use songs designed acoustically to degrade *as little* as possible, and therefore not as predictably, making the perceiver's ranging judgments more difficult. The logic of natural selection would suggest this, and there is a way to check it out in the field.

If singers use songs that degrade as little as possible, we should find songs degrading less in their original, native habitats than in some marginal or foreign habitat. Conversely, if the songs are "designed" to degrade predictably, degradation should occur predictably in any habitat either foreign or native to the song's evolutionary origin. In Maryland, Carolina wrens typically live in deciduous forests, but in Florida they are to be found in a quite different habitat—palmetto hummocks. Recordings were made of both populations and then played back to the birds both in their native and in foreign habitat. Sophisticated accoustical analyses were made, and the upshot was that Carolina wren songs are physically structured—the bird forms the notes—to degrade as little as possible. The songs recorded in Florida changed noticeably less in their native Florida hummocks than the same songs did when played through the Maryland deciduous habitat. And vice versa. Rather than being "designed" to inform threatening males accurately of their distance from the singers, *the songs hide this as much as possible.*

From a perceiving wren's standpoint, then, hearing a conspecific song is like listening to the sound of an incoming mortar round. It is crucial to become a good estimator of distance, to read through the "disguise," lest a great deal of time and energy be wasted in unnecessary investigations, and the cessation of foraging. And previous experiments had shown that Carolina wren perceivers *are* capable, at least up to a point, of making accurate assessments of distance based on degradation, in spite of apparent efforts by singers to make such assessments as difficult as possible.

We have come a considerable distance from the principles involved at the Krebs-Dawkins dinner party. It seems that we have a subtly different sort of arms race on our hands, at least among Carolina wrens. The long-distance communications, far from being simple and stereotyped so as to be perfectly clear, appear to be carefully structured to be confusing.

THE RANGING HYPOTHESIS

Were these findings from a single species of songbird applicable to others? If so, would they provide a synthesis of proximate and ultimate considerations, a linking of signal form and biological function, even tying in ecological and demographic elements which are all presumably related features of the evolution of bird song? Such questions led to what is called the Ranging Hypothesis, which sheds a coherent (if, from the standpoint of the romantic, not altogether flattering) light on the overall nature of the music of the birds. But, as we have said, animals do what they do, not what we would prefer them to do: Our potent urge to see them doing what we prefer can be illustrated by the way scientists traditionally used spectrograms. In recording bird song to produce a spectrogram for analysis, it was generally felt that a "good" spectrogram was made from as pure and undegraded a signal as could be arranged, a signal without any of the fuzzy outlines on paper caused by reverberation and other disturbances. In essence, you put the microphone as close as you could to the bird. Often, to refine their data, scientists would trace the spectrogram in black ink so as to remove any "noise," which was considered irrelevant. Of course such an ideal spectrogram represented only the sound as it was emitted from the bird's mouth, or perhaps as it was heard by the singer itself or another bird only inches away. On the other hand, the Ranging Hypothesis begins with the fact that it is the *degraded* version that is perceived in actuality by the birds for whom the sound was intended.

Bird song is, of course, only one form of long-distance signaling in nature. But *learning* long-distance signals appears to be restricted to only a few groups of birds— some of the passerines, hummingbirds, and perhaps a single species of toucan, those silly, beautiful tropical birds with huge colorful bills, that are related to woodpeckers. Evidently mammals, with the exception of humans and possibly some whales, do not actually learn their long-distance communications; instead they are programmed genetically. Parrots in the wild learn signals, but not for long-distance use. Among the cacophony of long-distance sounds that animals utter, learned ones have come about in very few instances, and this is an oddity that the Ranging Hypothesis must explain if it is to provide a coherent picture of this kind of animal communication.

Central to the Ranging Hypothesis is that perceivers

gauge (or range) distance from singers or other such callers by assessing the amount of degradation in the perceived signals. It stands to reason that perceivers can do this effectively *only* if they have the same signal structure in their own repertoire and thus in their own memory. Numerous experiments performed since those on the Carolina wrens—experiments with great tits and song sparrows, for example—demonstrate that the perceiving birds cannot use the song of another to range distance if the song is not part of the listener's own repertoire. How identical must the memorized signal be to the perceived signal? Are a few notes in a song syllable sufficient, or must the whole song be in the memory? No one knows yet. Nor do we know just how, neurologically, this judgment is accomplished. It might be similar to the way bats echolocate. It might be that the muscles of the syrinx respond in a nearly kinesthetic manner to the incoming signal—if it is familiar—and this response is somehow matched to the signal in the memory by some process wherein the bird converts a perceived song into the neural commands necessary to reproduce the same sound. Almost certainly, it involves some kind of time sense as well. Furthermore, it is probably no coincidence that birds' hearing is superior to most mammals *only* in this matter of peripheral hearing ability. (Our time-interval assessment takes place peripherally—in the ear—rather than in that portion of the brain given over to auditory matters.) Parakeets can resolve sounds separated by as little as one to two milliseconds; by contrast, humans lose sensitivity to sounds that happen faster than five to six milliseconds apart. Dogs fall somewhere in between.

Another basic question: What favors the development through evolution of long-distance signals in the first place? Clearly, dense vegetation and nocturnal life-style and other such factors that render sight and smell less effective. But life in the nonhuman world is nothing if not a subsistence life where food (energy input) and foraging, avoiding predators, and other basic matters often require a great deal of movement and thus energy output. Some animals build up capital reserves: Here one thinks of jays and squirrels hiding nuts against the thin days of late winter and early spring. In many animals, defense of a relatively large space in order to control vital resources is also part of the necessary daily round, and long-distance signals will be favored by natural selection whenever they can substitute for even a small amount of energy-demanding patrols on foot or wing. If such signals can lower the amount of energy use over time, the animal benefits. And the most complex use of long-distance signaling appears in *small* birds that have high metabolic rates, meaning small energy storage capacity relative to their daily energy needs. As noted, hummingbirds and other small birds like warblers are virtually on the edge of starvation by dawn. For such creatures, any energy conservation measure is of paramount advantage.

In this context, it is likely that long-distance signals evolved to control space by making an individual's presence *detectable* throughout that space. But a fairly simple noise, common to all members of a species, would theoretically be sufficient to be detectable. Why isn't it enough, evolutionarily, simply to say "Keep out! I'm here!" Why is some bird song, especially that of small birds, so complex? Another imputed role of bird song is to help keep species separate, to keep them from trying to breed across species lines. But here, too, a relatively simple and distinct species sound could suffice. You don't need to recite the Declaration of Independence to tell the man at Customs that you are a citizen of the United States.

The Ranging Hypothesis begins as just that—a hypothesis, a mental construct that attempts to explain such matters as those mentioned above. It tries to take into account such facts as: Small birds do the most complex singing; when a small bird is at the end of its rope, when its energy reserves are depleted (which is almost always), wasting further energy by responding to a nonthreatening bird is an act that is somewhere between bad and catastrophic. But if the Ranging Hypothesis is to be a scientifically sound hypothesis, it must do more than provide a plausible explanation for a few observed phenomena. It must also lead to predictions that can be tested or verified.

THE BIRDS' ARMS RACE

We start with an arms race over distance perception, and the separate roles of perceiver and signaler. The perceivers use some sort of distance-assessing mechanism (dare we call this unknown mechanism DAM for short?). By assessing distance, a perceiver is behaving in its own self-interest, avoiding wasteful activity. But if its DAM is too marvelously effective, and no matter what the signaler does he is always precisely read, song use would have no purpose (except to say "I am here"). It would not have needed to become as elaborate as it has. It would not be in the self-interest of the singer to waste a lot of extra energy on a complex song simply to announce its presence, and natural selection would logically eliminate elaborate song use or return it to a simpler form to match a simpler function. But if the perceiver is using his DAM in his self-interest, then the best way for the singer to overcome that and further his *own* interest would be to use the *same mechanism* to thwart him. Thus, the arms race.

In other words, a singer should use a song that *sounds close* to the recipient even though the singer may not be physically close. Such a song would have just the right quality of tone, pitch, and other accoustical properties to degrade as little as possible in the singer's local habitat, providing a narrower range of degradation by which the perceiver could make distance judgments. As we have seen, this is exactly the case in Carolina wrens.

But there is another way to sound close. The singer could thwart the perceiver's DAM by using a song that is *slightly* different from the listener's memorized songs—that is, new variations on a common theme. In this sense, the function of large repertoires of songs and high internal song complexity—songs composed of

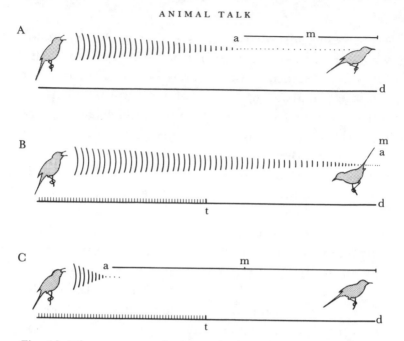

Fig. 13. The arms race happened when (A) listeners, on the right, began to range singers' songs by perceiving degradation and translating it into distance. Listeners ignored singers when they were far away and not a threat, even though the song carried all the way to *d,* past the listener. The listener would respond to the song only if the singer was at distance *a* or closer. In evolutionary response, singers developed means to have their songs reduce *m,* the distance between *a* and *d,* as depicted in the middle diagram (B). Singers might learn songs not in the listeners' repertoire so as to thwart the listeners' DAM, or they might produce songs that degrade as little as possible. Both mechanisms would increase *a* and get the listener's attention (or, dare we joke, to get the listener *mad* as in B). With *t* as a territorial boundary, B shows the singer's goal, whereas C depicts the listener's goal, to ignore the song unless the singer is a real threat to his territory. This is the arms race that produced the evolution of song-learning in small birds.

many time-ordered syllable types—can be understood best in relation to the perceiver's need to use its memorized songs as a critical component of its distance-assessing mechanism. And this gets us on the road to understanding why some birds sing so much more elaborately than others. It even leads us toward an understanding of the evolution of bird song.

Large repertoires and complex songs can be viewed as anti-DAM devices. They are false advertisements of where the singer is. If, as it seems, the perceiver somehow measures the song he hears against the song he has memorized and computes distance based on the difference, then a song that is close to his memory bank but not quite identical is bound to disturb the computation. Learning a new variation on a common theme would accomplish just that.

To recapitulate, in bird song we appear to have two different strategies at work—threat and disruption. The more straightforward strategy, threat, says "I am here and you know how close I am." Disruption decreases the *apparent* song distance. It says "I am here and I am close" when the singer, in actuality, is not close. Disruption can be practiced in either of two ways: (1) you sing your song in such a way that it degrades as little as possible given the accoustic nature of the environment, or (2) you vary your song just enough from the norm to be confusing. Either way, the singer can threaten a rival without the energy cost of getting near it.

■┃■┃■┃■

BORDER WARS

Even so, border disputes will inevitably happen and, in such cases, where still further proximity between rivals could lead to a truly energy-intensive battle, it would make sense if birds that had been faking their distance by messing around with the DAM of a rival used songs that could be accurately ranged—that is, if they used a song that was precisely stored in the memory of the rival. And that is a prediction that seems to check out. For that is just what we observe when two birds interact over boundaries: They switch from song types used earlier, before the dispute, to song types that match. This is commonly called "matched countersinging" and for a long time it didn't make a great deal of sense to observers. It was viewed as a form of escalation of threat, to be sure, but the reason for matching songs in this context was not clear.

This border-clash matching of song type can occur even in birds that have a single song type, like the Kentucky warbler. It turns out that each Kentucky warbler's version of the one song type is different, and each one sings with a different amount of emphasis (technically, a function of varying distribution of energy by frequency within the song's syllables). But a Kentucky warbler can change the allocation of energy within its song to match that of a rival or it can raise or lower the pitch of the entire song. By doing so, it is in effect saying "I am close and I know that you know it." On the other hand, a bird like the Carolina wren, which apparently cannot make

■┃■

such changes in energy allocation within a song, has taken the evolutionary route of multiple songs in an expanded repertoire. The average wren has over thirty songs.

FROM THREAT TO DISRUPTION

So much for matched countersinging at the frontier. Let's look again at the uses of long-distance singing. Some of it is, as we've seen, used as mere threat, and there is good evidence to show that the threats are more intense the richer the territory being defended. But why, in the course of evolution, would threatening behavior in some bird species have escalated beyond that into disruption, a different matter altogether? We need a plausible explanation of the conditions by which natural selection would have produced this more "sophisticated" behavior.

Now, an intense threat tells the rival how close he is to the defender and, by inference, how rich the defender's resources are. The rival can then judge whether the resources seem worth the risk of further invasion or not. Some border clashes will occur, but many will not. And that is that. There would be no selective pressures on escalating a nice honest threat mechanism into something more Machiavellian. But the situation changes if an individual can use a threat not just in defense *but to increase its own territory at the expense of the rival's territory.*

Remember: These animals have a low energy-storage

capacity relative to their daily energy need. The holder of a high-quality territory (one with lots of food) advertises this through high song output, a direct measure of the amount of time it is not looking for food. But if a singer uses its song also to disrupt rivals—even just a neighbor with a second-rate territory, forcing it to race around looking for a nearby threat instead of looking for food—then the singer may be able to *gain* some useful territory. This would be all the more effective if the disruption occurred during times when the neighbor's energy reserves are lowest. These moments occur in the daily cycle at dawn and in the early morning: the dawn song. It also occurs seasonally, as well as irregularly, as happens in the case of a sudden bout of extreme cold or a snowstorm.

In an experiment, during an abnormally cold period one January, Carolina wrens had food provided to them on alternate days. The cold had been greeted by a silence among the wrens, but when one of them was provided with food, it began to sing. Not only that, when it began to sing, its neighbors *stopped foraging altogether.* So the moral is "if you've got it flaunt it . . . and you may get more." And the upshot is that the rich *do* get richer.

For example, to turn yet again to the Carolina wren, in the deciduous forests of Maryland, winter is a truly terrible time for these birds, in particular when it snows. The snow blankets food sources and a wren must seek out such food as can be found in sheltered areas, beneath a fallen log, for instance. The birds are all literally on the brink of death for as long as the snow cover lasts, and the sad fact is that mortality among these birds in that area is often close to 90 percent. So the Carolina wren that can most successfully disrupt its neighbors, causing them to cease foraging at the most desparate time in their lives, is likely to inherit their territories . . . and any fallen logs there may be.

In only the rarest of cases—a few species—does bird society provide anything remotely similar to what we call a social safety net: Birds live in a universe that is different from ours in many ways, and it goes without saying that our standards of behavior in such matters are utterly unlike theirs. Most important, our choices are deliberate: Theirs are not, so ethical comparisons are irrelevant.

THE IMPORTANCE OF A GOOD ADDRESS

The nature of the neighborhood—not only its richness but its stability—seems to be an important component in the evolution of bird song. Most of the birds that do *not* learn their songs but inherit them live in areas where food supplies tend to be evenly distributed, perhaps rich, and where the local economy does not go through a drastic seasonal boom-and-bust syndrome. Such conditions prevail in the tropics, the characteristic haunt of what are called suboscine birds. These are perching birds that have more primitive songboxes, characterized by a few syringeal muscles—birds such as wood creepers, and, best known to North Americans, the flycatchers, a very large group of species in the tropics and a few of which breed in North America. One of these, the phoebe, has come to spend the entire year in North America, though it tends to move to warmer parts of the continent after the breeding season.

The songs of all these suboscine birds seem to be imprinted in the genes rather than learned: The songs are the same for every individual of the species regardless of geography. In fact, when a suboscine at one end of its species range sings a song slightly different from that of a member elsewhere, it is assumed that the new song is not a product of learning but that the bird may be of a different species. The last new North American bird species to be scientifically described was in 1960, when the Trail's flycatcher was split into two species—the willow and the alder flycatchers, a decision based initially on the difference between their songs.

In any event, among such birds in the tropics, there is a tendency for pairs to form monogamously and for life, and for competition for territory to be year-round—another form of neighborhood stability. In such a situation, mere detectability or, at most, simple threat would normally be sufficient to maintain one's place.

When, on an evolutionary time scale, neighbors compete in defense of well-established territorial borders, the economy of selection would logically favor the use of signal structures found in a neighbor's repertoire: the "I am here and you know it" lyric. To achieve the necessary level of detectability and threat, a simple repertoire—even a single song—would be sufficient. On the other hand, birds competing for unstable neighborhoods, places where one territory can be far superior to an adjoining one, or (in the mad scramble of northern spring) competing for a territory against a host of newly arriving conspecifics, the ability to disrupt one's new neighbors—as we saw with the Carolina wren—would become a selective advantage. Learning songs so that they become better adapted to the local neighborhood and its acoustical pecularities, in turn becoming less degraded as they pass through the area and reach intruders or neighbors, would be a distinct advantage. Such singers would *seem* nearer. And it is precisely in such places that you find birds, particularly those with low energy-savings accounts, which learn songs.

Thus what we call the Ranging Hypothesis provides coherent and consistent explanations for many of the observed phenomena of bird song—and the explanations tie this form of communication not only to bird ecology but to bird evolution, as well as shed light on some other puzzles of bird song.

■■■■■■■

REHEARSALS AND DUETS

One puzzle has long been the singing that goes on in late summer and early fall—sometimes called "practice," but often producing songs that the bird will in fact never use. Why would birds start singing again after the breeding season is complete and, in many cases, after territories have been abandoned or are no longer in need of such rigorous defense? (In these cases, there is a minor swelling of the sex organs at the same time—obviously not of itself particularly utile.) Some observers have suggested that at such times the adults sing in a kind of teaching mode from which fledglings can get their songs straight against the time the following spring when they will be needed. From the viewpoint of the Ranging Hypothesis, however, the need for more efficient DAMs suggests that more memory space be devoted to the business of learning songs—in effect, putting more songs in one's head—even if the songs themselves are never actually used. From the standpoint of natural selection, this view suggests that acquiring DAMs and producing songs are two separate entities. By singing a lot of songs to itself in the form of "subsongs," even though those particular versions may never be used in the "real" world of territorial defense, a bird may be programming its memory—salting away fodder for its DAMs. It may become a better perceiver.

Yet another puzzling phenomenon in bird song is duetting, in which a male and female produce alternating bursts of song. Far more common in the tropics than elsewhere, this was long thought to be an elaborate version of coordinated learning by a male and female to facilitate pair-bonding. But it has been shown among certain tropical wrens, for example, that the duets—which are typically very complicated—take place immediately and spontaneously during the first encounters between potential mates. There is no rehearsal period, no gradual coming together of a pair overcoming sexual reluctance and cementing relations in the joyous musical union finally achieved—which is what one would expect if the primary role of song was sexual. Instead, it appears that duets are *joint efforts at territorial defense*—the males singing to nearby males, and the females operating on alien females. That such duetting may also keep out potential cuckolders in the always tenuous monogamy of birds would seem to be a secondary role of the duet.

Overall, the Ranging Hypothesis suggests that song, and particularly song learning, is more a matter of competition among birds of the same sex rather than a matter of mate choice. From this perspective, mate choice is simply not necessary to explain what initially drove the evolution of song learning. It could have come to play an important role in sexual choice later, as would seem to be the case in such birds as marsh warblers and mockingbirds.

▐▐▐▐▐▐

VIVE LA DIFFÉRENCE!

In this regard, there is evidence that females for the most part cannot range male songs: They seem to have little neural space given over to song storage (an exception is in those species given to duetting). Without the songs stored in their minds, the females would have no way of gauging distance by the amount of acoustical degradation of a song, and this suggests that they perceive songs in different ways than males do. While territorial competition between males would favor many song traits that are not essential for species recognition, females may use the "backbone" song cues for species recognition. Wondrous variations on a theme probably mean little to them.

· Instead, once having recognized a male of her species, the female may attend to the male's *overall song output*— the amount of singing he does over time—as a "true" advertisement of the quality of his territory. If it's a lousy territory, he'll be too busy hunting for scraps of food to spend much time singing. A case in point, male American redstarts sing the same song repeatedly until they obtain a mate, when their richer repertoire of four to nine songs comes into play. Perhaps it is easier for a female to assess who has the richer territory if each male sings only a single song type. Another possibility— among mockingbirds and others that learn songs throughout life—is that the female may be cognizant of all the variations in repertoire or song-type and can judge that a male with a lot of variation is *older,* age being a function of ability to survive in a difficult world.

Further studies of the nervous systems of birds will shed more light on all this, but the Ranging Hypothesis suggests a test: If females cannot range song, it simply means that they can't tell exactly how far away a male is. But they can still hear him. As suggested by the redstart, it might do the male more good, when advertising the quality of his territory, if he were to repeat the same song type over and over. He would thus be assured that a female would associate him *and* the territory with a large amount of singing. Fantastic variations might confuse the female into thinking that there were several males present, each singing briefly. The test is to observe the singing in species with broad repertoires and see if the males use only one or a few song types before pairing, and a greater number afterward. If this is generally the case, then it would essentially disprove once and for all the idea that song repertoires arose through mate choice, and not for territorial defense.

▉▐▐▐▐▐▐

DIALECTS

Another tricky matter is local dialect. Bird dialects can be so distinct, even to human ears, that a practiced listener can tell from the song of a given white-crowned sparrow, for example, that it comes from one or another locale in California. (This is a very different matter than the variation in song type that can occur at the far ends of a species's overall range, variations that could be a step in an ongoing process of speciation. Instead, dialect is a term used to indicate a sudden change in song type or types that occurs often without any geographic separation between the two otherwise similar populations.) There are a variety of explanations as to how dialects evolved in the first place and what functions they serve. Many of these explanations are internally contradictory. It has been proposed that they might make sure that a population which has become well adapted to the fine differences in their particular habitat doesn't breed "out" at too great a rate. Another explanation is that dialects promote the rapid speciation of songbirds (after all, in terms of number of species, the passerines are supreme, accounting for nearly half of the 9,000-odd bird species and, the passerines being relatively recent in evolution, this must be a rapidly occurring phenomenon). Yet others have proposed that the value of learning dialects as a way of maintaining a well-adapted local population may explain the very origins of song learning itself.

The Ranging Hypothesis predicts that dialects will be most adaptive, most useful, when the straightforward signaling of distance—simple threat—is the primary function of singing, rather than the disruption of a rival's sense of distance. In such places, territory boundaries should be stable, defended year long by the same neighbors and with little between-territory variance in food availability, characteristics found in mild climates. Stress conditions, when it would be an advantage to flaunt one's resources, would not apply and disruption would not really work to a singer's energy advantage in such a place. And if this is the case, then perceivers in dialect species should respond strongly *only* to songs they can range. Songs from other dialects should evoke weaker responses: Only songs just like your own should create a strong response, because they would suggest proximity. A different song would mean that the singer is too far away to be bothered. On the other hand, in *non*-dialect species, the stronger response is to songs not in memory. In nondialect species, a bird would be unable to range an *unfamiliar* song and thus would need to react as if it might be close.

The geographic distribution of dialects supports this prediction from the Ranging Hypothesis. Eastern or northern populations of North American bird species do not have dialects, whereas populations of the same species located in warmer climates do have them. Red-winged blackbirds have one song shared by neighbors (i.e., a dialect) in California, but seven to nine songs in Wisconsin. While it is possible that dialects may serve as gene flow inhibitors (there is evidence to show that they don't), it would appear that any such functions are secondary to the role dialects play in the ecologically dependent business of territoriality. Thus, instead of being viewed as a core element to explain why and how song learning evolved, dialects should properly be viewed as merely one of the outcomes of natural selection operating on long-distance signals that are learned—and learned for other reasons.

This has been a long excursion into bird song, the most thoroughly and assiduously studied of long-distance signaling systems among animals. As far as is known, the Ranging Hypothesis applies only to birds. Birds, for example, share what is called an interaural pathway (a direct passage from one ear to the other) only with crocodilian reptiles, amphibians, and some insects. Mammals evolved without such a passageway. Just what part the interaural passageway plays in the estimation of distance is simply not known, but is likely to be complex. We humans may use the reverberation and intensity of sounds to make fairly accurate predictions of distance from the source of a sound, but it isn't known if these are especially accurate. It is simply unknown whether sound degradation plays any role at all among mammals.

It is instructive to learn that mammals (except maybe some whales) do not acquire their long-distance signals through learning, even though many are highly complex.

The haunting duets of gibbons, an amazing howling that begins at dawn, serve to defend territories, as do the songs of birds, but they are evidently based solely on genetics, not learned. The evidence for this is that hybrids among gibbons have hybrid calls—intermediate to those of their two parents.

Perhaps the chief lesson from this search through bird song, and of the explanatory power of the Ranging Hypothesis, is that it points out how a great variety of factors must be taken into account to explain anything so complex. We need to know not just what the signal is that an animal emits, but how it is perceived by the listener. We need to understand that natural selection probably works in different ways on the listener and on the signaler (though a single bird may be both at different moments). We need to look at the role of energetics in relationship to a host of factors, beginning with simple body size.

The entire business of long-distance signaling among animals has only recently been opened up for study, testing, and evaluation. It is no accident that birds have supplied science with the most grist for its mill in this particular arena: Bird song is almost ubiquitous and mostly diurnal. The Ranging Hypothesis seems for now to offer coherent answers to a number of puzzling phenomena in bird song and to offer testable questions that can be further asked of this most marvelous music of nature. That there will be other hypotheses, some perhaps more coherent and closer to a complete explanation, is something to be expected . . . and looked forward to. For that, happily, is the nature of science. But in the meantime, the Ranging Hypothesis highlights the fact that communication is a transaction between at least two individuals; and it explains that transaction in both the proximate and ultimate senses, and, in so doing, draws our attention to the crucial importance of the perceiver.

Seven

||||||||||||

LISTENING TO LION?

*In which the philosopher's lion does
not speak; information theory is dealt its
death blow; the role of the perceiver gets
its due; the matter is illustrated by the
lives of rattlesnakes and ground
squirrels; and the communicative abilities
of the young receive due credit.*

The philosopher Ludwig Wittgenstein said: "If a lion
could talk, we would not understand him." As with all
such pronouncements from the deep, this is a comment
that one can turn around and around in the mind, spin-
ning endless streams, even webs, of thought. For example, if a lion were to talk, what language would it speak?
Some form of Lion, one assumes, but would that be in
a language of visual images, of raw emotions rising from
some primeval depths, of olfactory paths, or what? In
fact, we wouldn't understand it, and the reason is that,
whatever language the lion talked in, it would not be

ours. We would not know how to listen. And because we wouldn't be capable of listening to the lion, the lion would be incapable of talking to us.

For a philosopher, this might seem a paradox, but for an evolutionary biologist it now makes perfect sense.

The facts are, of course, that lions have been speaking (vocalizing) in Lion for a long time, that they haven't been vocalizing in our direction or for our benefit, and that we do know a bit now about how to eavesdrop on them. Lion vocalizations have a fair amount in common with those of other animals and we are learning to understand them. And by now the reader should know that Wittgenstein's *obiter dictum* is a mind game: Lions will never "talk." What is important about Lion is that it is a system of communication that has arisen among lions and *it has accomplished a variety of things that are inextricable from the rest of the business of being a lion.* We can understand it to some degree by recognizing that Lion arose through what amounts to a series of rather draconian trial-and-error experiments that began with the utterances of some predinosaurian ancestors that split their time between land and water.

Our beginning understanding of Lion, or any other system of vocal communication among animals, is gained in the manner of a person observing others through a two-way mirror. And each time we have gained through the glass a glimmer of what the conversation/actions appear to represent, we have laid on a definition of the overall situation, or forced what we have seen into a concept that is at hand. The contention here is that the definitions and concepts have heretofore tended to obscure the phenomenon under study, as if the definition were scrawled boldly on the two-way mirror itself, thus obscuring the view, indeed even limiting vision altogether.

We have looked at least obliquely at a number of these definitions or assumptions and complained about them to varying degrees, at the same time showing two approaches to animal communication that proceeded without reference to these definitions but to real events and attributes in the real world. The Motivational/Structural rules that seem to govern short-distance communication among mammals and birds and derive from evolutionary roots in reptiles and amphibians also seem to beg the standard definitions of what animal communication *is*. Similarly, the Ranging Hypothesis that was offered as a means of understanding the nature of long-distance communication, at least among birds, raises yet other issues at odds with the standard definitions.

In the light of these two new hypotheses about animal communication, it seems time to look yet again at the other definitions that have been employed, in particular what has been called the Information Perspective. This may seem like flogging a dead horse, but the fact of the matter is that, among those who study animal communication, the horse is by no means dead yet.

IIIIIIII

THE INFORMATION IMPEDIMENT

Appeals to the idea of "information" tend to include (and even confuse) both information as generalized "meaning" and information as something more mathematical—bits of information that can be employed to reduce uncertainty in quantifiable ways. Such appeals, especially when combined with analogies to information-processing systems such as computers, appear to have provided a rigorous approach to one of the most intractable problems in biology—and indeed in philosophy: the origins, development, and maintenance of living organizations.

We live in a world characterized by what the physicists call entropy—the tendency of energy to dissipate. An example of entropy: using a blowtorch to heat up a bathtub of water. The energy, which is highly concentrated in the blowtorch fuel, is transmitted to the water in the form of heat. Basically the same amount of energy will still exist, but most of it will have spread throughout a great deal of water. It is now in a more dissipated form. If you then tried to use the warm tub of water to heat up yet another tub of water by mixing the two, the result would be twice as much water that was half as warm. That is the nature of entropy, and entropy seems to govern most of the phenomena of the world. The existence of life, taking the dissipating resources of the world and turning them, contrarily, into little packets of organization that grow and reproduce and in some cases move on their own is an (at least temporary) anti-entropy phenomenon and requires explanation. "Information" seems to point toward an explanation. It seems to provide another "something" that operates in an organizing manner in the phenomena of the world.

As psychologist Susan Oyama wrote recently:

> In an increasingly technological, computerized world, information is a prime commodity, and when it is used in biological theorizing it is granted a kind of atomistic autonomy as it moves from place to place, is gathered, stored, imprinted and translated. . . . Information, the modern source of form, is seen to reside in molecules, cells, tissues, "the environment," often latent but usually potent, allowing these entities to recognize, select and instruct each other . . . to regulate, control, induce, direct, and determine events of all kinds. When something marvelous happens . . . the question is always "Where did the information come from?"

To perceive information in a molecule may seem a bit abstract, a bit far from our daily ken, but to perceive information in something as similar to our own behavior as the vocalizations of animals seems altogether reasonable on the surface, especially in light of the centrality of information in explaining the weird but wonderful habit of life in general to oppose the inexorable, dreary rule of entropy. Life is organization, and organizations survive by means of the flow of information. Carried to the extreme, this idea provides an image of life as a great gyre of interoffice memoranda. Listen:

"Communication can be defined as the conveyance of information from one organism to others."

"Communication consists of the transmission of information from one animal to another. Information is encoded by one individual into a signal. When received by another animal, this information undergoes decoding . . ."

"The informational perspective is concerned with the mechanisms as well as the functional significance of communicating. It treats formalized signaling (i.e., behavior specialized to function by making information available . . .) as only one of many sources of information in a complex process."

What is conceptually implicit in such definitions is, among other things, that the coding, transmission, and decoding of the information in animal communication is equally beneficial to the sender and the receiver—and beneficial in the same general way, for example, providing the proper organization of animal relations or society. In such a *mutualistic* system, the only way one animal gains at another's expense is by withholding some or all information or by outright deceit. Since biologists occasionally remind themselves that natural selection does not work at the species level but instead chiefly at the level of the individual—that genes are, in a metaphorical sense, selfish—then a search had to be mounted for widespread examples of deceit in nature. And examples, though mostly anecdotal, began to turn up through the two-way mirror. Also implicit in the Information Perspective is the concept of the listener as basically passive. Upon decoding the information delivered from the sender, the listener knows what is on the sender's mind (or, in the case of deception, what the sender wants the listener to think is on its mind) and reacts appropriately.

One conspicuous exception to the widespread enthusiasm for this way of looking at animal communication was to be seen, again, in the work of Krebs and Dawkins. They looked at both the direct physical action an animal can take (like shoving or hammering) and at an animal's signaling as sharing a pragmatic quality—a quality that does not require informational thinking. They called this activity "manipulation," the admittedly negative or cynical connotations of the word helping to emphasize the fact that most animal behavior appears to bear a component of self-interest that can, to a degree, be to the *potential detriment* of other communicants. To manipulation, which they emphasized was metaphorical—that is, a way of thinking about such transactions—they added the complementary role of "mind reading." These two roles, which can of course exist virtually simultaneously in one animal, provided the listener with a somewhat more active role—and a purely self-interested one at that—and this seemed to be in better accord with what is taken to be the logic of natural selection.

But there were problems with this formulation, too. They arose from the distinction between proximate and ultimate effects. Manipulation, critics pointed out, really referred to the overall—the average—impact of the signal over time on the long-range fitness of the signaler and the receiver. But the immediate (proximate) coupling between the act of signaling and the reactions to the situation created by the signal still depended on the transmission and decoding of something called information, information that *caused* something to happen. To a passing female, the rutting bull says, in effect: "I may look aggressive, but I'm really looking for a mate." And the female, reading this *information* embodied in what could be a very complicated array of signals from scent and body language to a slightly different vocalization than standard hostility, says in reply (by some equally complex array): "Okay, I read you but I'm not sure I'm interested." and so forth. Information exchange. And also, as we say, a considerable emphasis on deceit of some degree: withholding information, or misusing the information channels and signals.

Theoretical and experimental studies continued, debating whether withholding information or issuing false information was the most conducive strategy to overall (ultimate) fitness or survivability. A conclusion arose that withholding information is the best thing to do, since outright deceit is easier to catch. This is a bit like discussing the tactics employed by angels, since it is by no means clear that either of these tactics were actually employed by animals. By this time, information was a separate, operating player in the drama. What had begun as a vague bit of shorthand ("information . . . well, you know what I mean") had left the realm of descriptive metaphor and had become a *thing,* indeed a causal factor. Until it is returned to its use as shorthand, it will continue to impede our vision.

PTOLEMY'S MACHINE AND THE PERCEIVER

We have made it a practice so far to use the word "per-ceiver" in these discussions, and not "listener" or "hearer" or "reactor." And that is because the only way to understand what a signaler is up to, what it can accomplish, is to understand the extent of the perceiver's extremely active control over the situation in which the signal is uttered. This is a largely new emphasis in the science of animal communication.

For example, we have looked at songbirds in the light of the Ranging Hypothesis. One could, by a stretch of imagination, say that a songbird carrying on with a learned variety of songs is encoding information into them to be conveyed to a putative neighbor. By producing a song that is not in the repertoire of the neighbor, the singer is providing a kind of misinformation: "Do you hear this, pal? You don't know where I am, do you, so you better jump up and down and fret as if I were right on your doorstep." Thus the singer brings confusion to his enemies, creating a false confrontation that would otherwise take a lot of actual patrolling of borders or aggressive acts: In the confusion, he may even gain materially in one way or another. By this stretch of the imagination, the bird is producing territorial "information" (or dis-information). But the song appears to have another function, an apparently ancillary one: Females are attracted or repelled by it. At the same time, we have seen that females typically do not possess the neural equipment for ranging distance by the amount of degradation of the structure of the sounds emitted as they pass through the environment (the "information" putatively encoded in the song). Instead they hear it in a different way. A female may actually hear the marvels of the male's song, its variety, its "information" but she evidently gets nothing out of it. Instead she hears it as a lot of singing or a little singing or an amount somewhere in between. Females appear to be interested in the amount of time a male takes away from his foraging time to sing. The more singing he can produce, the more food-rich his territory is. (Experiments have shown that when food is added, male singing increases in duration. Q.E.D.)

If we continue to look at this complex communication as the exchange of information in some formal sense, we have to assume that the singer is encoding two totally different kinds of information into his song. One relates to distance and is a message to males. The other is about the richness of his territory, and is intended for females. One is "dishonest" and the other is "honest." The message encoded for females functions honestly to provide an accurate statement of resource richness—a classic case of reducing uncertainty. Meanwhile number one operates in an opposite manner: Far from reducing uncertainty, it increases it. In emitting a song that cannot be accurately ranged by a neighbor, the singer is indicating only one thing that could in fact be thought of as information, and that is that he is somewhere in the vicinity. The remaining import of the song is not information ("I am nearby") nor is it really misinformation ("I sound close") but really information's true opposite, which is no information ("I am somewhere *but you have no idea how close or distant"*). Far from reducing uncertainty or creating a false uncertainty, the song serves to increase overall uncertainty, and in the face of that uncertainty, the listening bird has no choice but to act as if the worst were at hand.

It would, on reflection, take an amazing verbal skill for a human speaker to produce a sentence or paragraph that had two totally different forms of information embodied in them, each of which would produce a totally different desired reaction on the part of each of two simultaneous listeners. Such an onus seems too great to place on a bird that weighs some fifteen grams, even one that has occurred as a product of a long and complex evolution. It is in fact what might be called Ptolemaic.

One of the Ptolemaic astronomers set out to produce a "machine" that would reflect the motion of the heavenly bodies. But the machine had to be made impossibly intricate because Ptolemy, like virtually everyone at the time, believed that everything in the sky revolved around the Earth. So, in order to handle the erratic apparent motion of planets, he had to add epicycles to his cycles, and to the epicycles he had to add yet other epicycles. The machine *worked* fairly well: It just happened to be ludicrously complex, which might have been seen as a clue that it was also *wrong.* The Information Perspective in animal communication has become a bit like Ptolemy's machine.

An alternate way to think about these matters is to return to the stickleback fish, which get furious upon seeing red—be it another fish or a truck. Red triggers aggression, and over the long haul of stickleback evolution, those that have been so triggered have survived to produce viable progeny: That is, in the matter of inter-male hostility, they are fit. It is simply what the males do and over the long run it works. Similarly, singing a variety of songs is what some male birds simply do, the reason being that this is what their ancestors did, and it worked. The ability to learn new songs is an inherited ability, even if the actual song types vary somewhat from bird to ancestor bird. A new variation in the form of the song acts in such a way as to foul up a neighbor's distance assessment, and this in turn triggers certain behavior by the neighbor that is, over the lifetime of the singer, to his ultimate advantage.

One does not need to infer what we generally mean by *content* in such communication. Instead, one need only think of it as *form*. A change in form—in this case, acoustic form—causes a functional difference in both the immediate neighborhood and in the proximate time sense. Thus the song can be looked upon as a means that has ultimately helped the bird to *influence* its social environment. Even this is an overstatement. The signal itself doesn't make anything happen. It becomes instead part of the overall environment which the perceiver is constantly assessing, as is the signaler. Functionally speaking, the signaler is managing the behavior of perceivers only to the extent that the signal alters the immediate environment to which the perceiver is constantly reacting in more or less statistically predictable ways.

Similarly, in the matter of females listening to that same song, acoustical form is involved merely to identify the singer as a member of her own species (which is to say, an okay potential mate). But what sets her off, what triggers her positive or negative response, is not so much any further aspect of acoustical form in the song but merely how long it lasts. She responds to comparative length because her mother did the same thing and it worked. All three—the singer, the female, and the male neighbor—are simply acting in their quite different individual interests.

An important word in the last sentence is "acting." The female actively moves around, hearing the songs of many males, in effect assessing them one against the other comparatively, based on a sense of time. She is assessing her environment, of which the male song is perhaps the most important ingredient during the breeding season. Her subsequent actions—those that serve to determine her choice of mate and then to behave in such a way as to complete the breeding process with him—also enables her to manage her environment. The male neighbor, meanwhile, is actively assessing *his* environment, of which one major component is the richness and security of his territory and another is the sounds emitted by potential invaders.

So the bird's song functions on two quite different levels. On the level that concerns the female, it functions largely within the constraints on the singer of food resources, which is an energy constraint. As a territorial proclamation, the song functions within a large number of constraints, including the acoustical properties of the nearby environment, the size of the bird, and the availability of food resources. But a major, and often overlooked, constraint is the evolved ability of the neighbor to assess the signal. The active role of the perceiver—for example, the development over time in some birds of distance-assessment mechanisms (DAMs)—could well be part of an arms race that has, in a very active way, led to the development of elaborate antiDAM song repertoires. In this sense the perceiver, by actively assessing its environment, including the vocal signals that are a part of it, and acting in its own interests as a result of that assessment, has shaped the very nature of those vocal signals. This is like the audience's role in determining the routine of the stand-up comic, and we might continue that analogy a bit further.

The comic would be in a nearly impossible situation if he or she faced an audience made up of equal parts of sophisticated adults (the sort that read *The New Yorker* mostly for its witty cartoons) and ten-year-olds. The comedian is almost certainly going to lose half the audience, since most kids don't find *New Yorker*-type humor funny at all, and most such adults cringe at what a ten-year-old finds hysterical. Parents put up with the awful jokes that make kids laugh, assuming that they will grow up one day and learn to appreciate more refined humor: that is, adult humor. But what does this have to do with animal communication?

Scientists who have perceived a high degree of semanticity in the likes of vervet alarm calls have also noted that young vervets get these calls wrong at first. A neophyte vervet, upon hearing an alarm call, tends to look at its mother to see what she does, and then follows suit. Similarly, an adult vervet won't pay any attention to a "wrong" alarm call by a neophyte, such as the "eagle" call used when one is alarmed by a falling leaf. An adult, similarly, will largely ignore the alarm call a young vervet makes when it has seen a baboon. The assumption is that all the calls of young vervets are imperfect and need to be refined through learning and experience before they achieve the fine semanticity of the adult usage. This assumption is based on the implicit notion that there is specific information embodied within the utterance that properly refers to specific objects or events in an abstract manner. In other words, the dumb kids shriek "eagle" at anything overhead because they don't yet understand the true meaning of the word.

▌▌▌▌▌▌

PRAIRIE DOGS AND RATTLESNAKES TO THE RESCUE

Some elaborate studies of ground squirrels and their relatives, and their reactions to predators such as snakes, not only sheds a new light on this youthful "ignorance" but also furthers our understanding of the nature of the transaction that goes on in animal communication. It had been discovered that ground squirrels—highly social little rodents which are often pictured standing up on their hind feet looking alertly about—have a variety of alarm calls that seem to match the identity of a potential predator with a semanticity not unlike that of vervet monkeys. Faced with an aerial predator, a ground squirrel emits high-pitched whistles consisting of a single note. On the other hand, a mammalian predator such as a badger elicits a different call—a segmented call that is lower pitched.

These rodents seem to be discriminating even further—vocally signaling the difference between raptorial birds and vultures, and even between harriers (low-flying hawks that can suddenly appear from over a nearby rise) and high-flying hawks. In addition, the alarm call warning of a coyote or dog sends the nearby ground squirrels down the nearest burrow, while the call signifying *badger* sends them to the nearest burrow *that has a back door.* Badgers are, of course, burrowers themselves.

Over the years, Donald H. Owings of the psychology department of the University of California at Davis and various colleagues have been looking more closely at the nature of these calls, the reactions they elicit, and the process by which young ground squirrels get up to speed in the alarm business. In particular, he organized an elaborate experiment that concerned black-tailed prairie dogs and their reactions to intruding snakes.

These rodents, which are akin to ground squirrels, tend to be aggressive when they find themselves in the presence of snakes, sometimes biting or slapping at them with their front paws. They also execute what are called "jump-yips": The animal simultaneously leaps up in the air, its body vertical, and lands on all fours while emitting a sharp cry described onomatopoetically as a "yip." Occasionally in such circumstances the prairie dog will emit a bark, a shorter, higher-pitched sound that is not usually accompanied by a jump. The experimenters tormented their prairie dogs by exposing them to large rattlesnakes, small rattlesnakes, large bull snakes and small ones. The rodents' reactions were carefully logged and analyzed, and it became clear that the adults made vocal distinctions between large and small rattlesnakes and between rattlesnakes and bull snakes, but it didn't make any difference to them if the bull snakes were large or small.

Now, if one thinks of the presence of a snake among prairie dogs as presenting a situation of endangerment, and one thinks of the reactions of the prairie dogs as determined by their assessment of endangerment, then the logic of these reactions is quite clear. An adult prairie dog presumably has some resistance, thanks to body size, to the effects of rattler venom—and the bigger the rattler, the more venom it has, so the less resistance the prairie dog has. The more frantic the jump-yipping, the more likely is the prairie dog to attract others of its kind in a mobbing action, the better to get the snake to move away. The smaller the snake, the less necessity is there to enlist the help of others. At the same time, it appears that a bull snake of any size poses little threat to an adult prairie dog. The jump-yipping in such cases is more restrained, its form and pattern signaling that there is less endangerment and that the other prairie dogs can go about their business while one or a few monitor the snake with an occasional yip.

But the reactions of young prairie dogs is quite different. They make no distinction between the size of rattlesnakes—they respond uniformly to all rattlers with barks, high-pitched sounds of acute alarm. On the other hand, they do distinguish between large and small bull snakes, a small one evidently posing little threat. The alarm calls of the young—a gradation going from barks of special alarm through barklike yips to low-alarm jump-yips—is much the same as the adult alarm system, only it is differently applied. But is it inappropriate or misguided? Owings and his collegues suggested that the young prairie dog's alarm system was precisely adapted to the kind of danger it was in.

A young prairie dog, being nowhere near big enough to withstand any venomous attack, would not need to distinguish rattlers by size. They are all equally lethal. Best to start barking frantically, which will serve to call in a lot of adults in response to a sound warning them of great danger. Rattlers, like most snakes, have poor hearing: A bark is a more intense and audible sound, but it is not accompanied by jumping, which would permit the snake to get a good visual fix on a possible victim's whereabouts. Part of a young one's assessment of the danger it is in is the probability that it is not yet agile enough to engage in the harassing attacks that are often associated with the jump-yip display. The reaction of a young prairie dog to a large bull snake is somewhere between a jump-yip and a bark, since a large bull snake might prey on a young prairie dog. Finally, a small bull snake poses little threat and gets the youthful jump-yip treatment. Just as awful jokes that send ten-year-olds into hysterics are just the right thing for a kid's developing sense of humor, the alarm calls of young prairie dogs are suitable responses to the situation in which they find themselves. In a way, this makes perfectly logical sense: They use what amounts to a textbook example of the

Motivational-Structural rules, and, by doing so, tend to elicit the needed behavior on the part of the adults, bringing them in as protectors . . . as needed.

It is interesting, in this connection, to note that parents of young prairie dogs, upon seeing a bull snake, do not send forth a signal bearing information that would be of specific value to their own young. They don't say: "Watch it, junior, wherever you are, I've got a BIG bull snake here." They simply react as adults do to bull snakes in general. It is up to the young to assess that reaction and take whatever action is called for.

Similarly, young vervets have an alarm they emit upon seeing a baboon that has ventured too close. Young vervets are endangered by baboons, adult vervets are not, and not only do the adults ignore baboon alarms by young vervets but they do not take the trouble to warn the young in a troop that there is a baboon nearby. It's up to the young to notice such things. And typically, they do. Similarly, what seems to be a misreading of a predator threat by a young vervet, say, giving the leopard alarm at the approach of any large creature like a stork or a warthog, may not be mere juvenile ignorance, but a reaction to a very reasonable threat.

Thinking about vervet calls or prairie dog jump-yips as semantically loaded signals bearing specific information tends to make one think of juvenile utterances merely as ill-developed versions of the perfection present in adult utterances. It puts a far greater emphasis on a kind of learning process that smacks very much of the kind of learning that goes on when a human child learns to read, messing up, gradually coming to associate letters and syllables and words with their proper meaning. This is in fact what is called "culture" by anthropologists, the learning that goes on quite separate from any biological considerations except the inherited capacity to learn.

Any human baby is genetically disposed to learn language. Which language it learns is utterly arbitrary, usually depending on what language its parents happen to speak. The learning of alarm calls by vervets and prairie dogs, when considered as a series of lessons in semantics, suggests that these animals learn culturally, as we do. The same has been suggested for song-learning in birds. There was a time when a great deal of effort was made to look into the vocalizations of spider monkeys, for example, in the hope of obtaining some crucial information about the way humans learn to use language. As one of the original investigators has recently said, "What was learned was nothing about human learning but a great deal about the vocalizations of spider monkeys."

In short, studies like that of the prairie dogs and the snakes suggest that "cultural" learning is not the way of nonhuman animals, as it is of us humans. Their use of communication signals is firmly embedded in the constraints of biology. The signals have not evolved to exercise the kind of control over passive listeners that is implicit in an information perspective on such matters; instead they have come about to modulate the immediate environment—to reduce the difference between the situation that exists at any given time and a situation that would be better for the signaler. The signal operates within a series of constraints that comprise the "situation." The signal is a means by which to *keep the attention* of listeners and even in some cases keep what might be thought of as their *favor,* their benevolent attention.

THE CONSTRAINED SIGNALER

The rattler's rattle may signal its state of mind—a certain amount of irritation and preparedness to strike—but it is also a direct semaphore of its actual size. And, too, the sound of a particular snake's rattle is directly related to air temperature: a cold snake's rattle simply sounds different than if it were warm, and a cold snake is less able to move as swiftly as a warm one. The sound of the rattle signal is thus constrained by circumstances quite beyond the snake's control and quite obviously beyond any intent one might attribute to it to convey information. By producing this particular signal, the snake is expressing its state of mind in regard to the situation at hand. It does so because that is what its progenitors have done in similar circumstances, usually being successful in changing those circumstances to more favorable ones. The snake is not deliberately encoding information into its rattle: In fact, it has very little control over it, and none whatsoever over an important component of it (or constraint on it), which is the temperature of the surrounding air.

The rattle is best seen as a part of the snake's repertoire of behavior designed to manage his situation. To the degree that there is anything in the signal that we might, metaphorically, call "information," it is in the activity of the perceiver—in this case, the prairie dog's assessment of the degree of danger presented by the snake. In this assessment, the signal itself, as modified by the temperature, is important; so too are such matters as the size of the snake. The prairie dog assesses the totality of the situation and then proceeds to act in its own best interests, to manage *its* situation. It may accomplish this by shifting from perceiver to signaler, and barking.

It is basically impossible to separate the bark response of a young prairie dog from the rest of its physical activity, and to consider the bark itself something uniquely bearing information. The bark is part and parcel of the total immediate situation, which includes such factors as the young prairie dog's size, its state of mind, the danger of jumping up and down and thus being visible, and the need for help from adults. Similarly, the ground squirrel's badger call is part and parcel of the behavior associated with it—the search for a burrow with two entrances. It is not apparently a matter of a ground squirrel emitting the badger call and then, upon some sort of reflection, ignoring a one-entrance burrow to find a more suitable one. The two actions—call and flight—are essentially one reaction to one environmental stimulus or more.

And it is a sure bet that, among vervets, merely seeing one's fellows head for the treetops would be enough to get a deaf vervet to do the same. That the leopard alarm and the flight to the treetops are associated would seem to be less a matter of semantic information embodied in an otherwise abstract noise and more a matter of both the call and the flight being inseparable aspects of the same thing: a unity of form and function—what might be thought of as an effective multipronged antipredator package.

At this point, one might ask why a vocal signal would have come about as part of what amounts to physical flight behavior. Is there some sort of systemic altruism involved? Studies of ground squirrels have suggested that their "alarm" whistle operates very much to the signaler's advantage. It reduces the conspicuousness of the caller, masking its sprint for refuge by evoking a great deal of running around by other ground squirrels. This is as good an example as any of the self-interested management by the signaler and the self-interested assessment by perceivers.

MANAGEMENT AND ASSESSMENT

Is this talk about management and assessment, as opposed to manipulation and mind reading, a kind of nitpicking, or does it provide us with an important new view of what is actually going on in the vocal communication systems of animals?

If you perceive the central question about animal communicative behavior as what it serves to accomplish rather than what it symbolizes, you are led to a somewhat different view of deception among animals. Krebs and Dawkins brought up the idea of an animal exploiting the essential "lawfulness" of things in nature, be it a rock or a fellow animal's behavior. In the conceptual scheme expressed here, where an active perceiver is essentially in greater control of events than had previously been credited, the concept of lawfulness in both the structure and the way in which signals are used is seen as representing the constraints of the situation at hand. And this is different from the existence of a "book" of set rules to be followed or violated with humanlike duplicity—a deliberate flouting of the regulations that define honesty. Instead, if you think of the category of constraints most relevant to what is usually called deception as *the expectations of others,* then "deception" is merely another category of actions that serves to circumvent the constraints of the communicative situation at hand. It might be more useful—particularly as a tool for analyzing the complex goings-on of animal behavior—to think of what has been called deception as one degree or another of *action that violates the expectations of others.*

We mentioned that some birds, among other animals, employ a variety of threat displays, often using one until it diminishes in power to warn of attack. At that point the animal may switch to another display. No one would call this switch *deception.* It is, however, behavior that violates the current expectancies of the perceiver. So what we rather arbitrarily call deception on the part of animals might better be thought of as a series of graded behaviors along a continuum. At one end of the continuum is something as simple as a changing threat display; at the other is the use of a vocal signal in an unusal manner to fit an unusual context.

In summary, in this overall interpretation of the actual events involved in animal vocal communication, we have the signaler employing a vocal sound that expresses its state of mind: an attempt to modulate its immediate environment. And of course, this vocal signal—part of a greater complex of behavior on the part of the animal in most instances—has come into existence because it has in fact functioned over hundreds or thousands of generations in some form close to what the animal uses today. The vocal signal has to function within the constraints of the environment and the immediate situation. And one of the major constraints in the environment is the manner in which others perceive the signal and respond to the new situation.

Meanwhile, the perceiver is at the same time seeking to modulate its own situation and react to its immediate environment to further its own interests. Its reaction to the signal emitted combines not only the signal and the associated behavior of the signaler (if this is perceptible) but also a kind of statistical probability that the signal indicates a narrow range of events that are about to happen. (If the dog draws back its lips, there is a probability worth noting that it will bite.) In this sense, the animal signaler, like the stand-up comic, is obliged to "learn" (or somehow to know) how the perceiver will react. This social feedback from perceivers to signalers is highlighted by a study of cowbirds. When reared to adulthood in isolation, cowbird males sing the cowbird song far more potently than do wild males, which of course are in fairly regular contact with each other. This is puzzling until one realizes that isolation brings not only a lack of society but a lack of the normal social consequences of singing. All young males are naturally disposed to sing potent songs, but such songs typically evoke aggression from dominant males—that is, trouble. So a young male cowbird learns from his perceivers to modulate his voice.

The active perceiver, then, and not a passive listener, can be seen as the chief constraint on the signaler. The perceiver, in this sense, becomes as much a controller of the proximate situation as the signaler—perhaps even more than an equal. It could be said that the perceiver creates the final message. Ultimately the perceiver can be seen as the engine that drives the evolution of animal communication.

Animal vocalization can now be understood as embedded in the overall behavior of the animal signaler *and* the animal perceiver, along with such matters as their individual ability or inability to learn, and whether to take or not to take a step in a continuing evolutionary arms race. In this sense, animal vocalization is more than just the activity of individuals and the situations each finds itself caught up in. It can instead be understood as an ongoing interindividual *patterning* of behavior—behavior that can be understood both in its immediate effects on the lives of the participants, and as a system that arose, as does every other aspect of an animal, via the logic of evolution.

It is not necessary to tangle oneself up in the semantic problems that adhere to such words as *language, information, linguistics, culture, word, deceit.* These are purely human terms, abstractions in search of meaning. Animal communication just isn't that complicated. Nor is it that simple. It is always a matter of urgency. It is of the moment, of the entire situation that confronts an animal, and thus always "important." Animals in nature do not engage in small talk, or in any form of talk at all.

Eight

||||||||

THE ORIGINS OF SPEECH

*I*n which Hoover speaks, if tipsily;
the distinction between science and
speculation is made; and the authors go
on to speculate anyway.

In November 1978, it became clear that Hoover was speaking English. He was seven years old and for several years he had grown more and more vocal. An observer then noted that "he says 'Hoover' in plain English. I have witnesses." Hoover was rewarded when he was overheard talking, and before long, without any verbal prompting from his caretakers, he would rear up and say things like: "Hoover, come over here, haw, haw, haw." He sounded a bit inebriated. And he had a Boston accent. No hard *r*'s.

There would be nothing remarkable about this except

for the fact that Hoover was a harbor seal, a denizen of the New England Aquarium. There was speculation that Hoover was learning his raspy English phrases from a putative wino, who frequented the aquarium in his daily meanderings around town. And while the wino might well have taken it for granted that a harbor seal would become an intimate interlocutor, something perhaps like Harvey the six-foot rabbit, there was a ripple of amazement in scientific circles since mammals, with the exception of dolphins and a few apes, were not supposed to be able to imitate human speech.

Scientists from the National Zoological Park in Washington, D.C., were dispatched to Boston and tape-recorded Hoover's utterances, converting them to spectrograms. These were matched against a human's imitation of Hoover's imitation of the human voice—in fact, the senior author's imitation—and though the two tapes sounded alike, the spectrograms looked quite different. The fact is that even a spectrogram of a human voice uttering the same vowel sound months apart may look quite different: The way a sound is perceived, it turns out, is as important as the sound's physical characteristics. Research on human perception has identified the essential acoustic *patterns* of speech that let a person perceive a given word or syllable correctly. These essential stimuli are called speech cues. A broad *O* may be heard as such only relative to the pronunciation of other vowel sounds. In other words, what is one man's *oooh* may be another's *O,* depending on context. Hoover's speechlike sounds contained enough of these cues, particularly in the lower frequencies, to be heard as familiar words. In any event, Hoover's largely unique capacity was judged to be an indication that harbor seals' vocalizations—mainly a male affair—reflect the capacity to mimic a local dialect among seals and the matter was mostly forgotten except as an amusing curiosity.

∎∎∎∎∎

MIMICRY

The inability of most mammal species to mimic human speech has, of course, been frustrating to scientists, who have sought to set up direct communication with animals. Primatologists have had to teach chimpanzees other languages, such as sign language, or the manipulation of colored objects or computer keys, to look into their ability to cope with words and sentences. The best-known mimics of human speech are parrots, which, in nature, tend to live in large, often noisy flocks. Mating pairs of parrots are known to mimic each other's calls, presumably the better to identify each other in the crowd, but their ability to mimic human speech patterns has delighted and amazed people throughout recorded human history. And in the past decade, at least one, an African gray parrot named Alex, has been the subject of scientific training and study, learning to use his version of English words in what appears to be direct communication with his handlers. He can, says Irene Pepperberg of the University of Arizona, use English vocalizations to "request, refuse, identify, categorize, or quantify more than 50 items. . . . He has acquired functional use of 'no' and phrases such as 'Come here,' 'I want X,' and 'wanna go Y' in order to influence (albeit to a limited extent) the behavior of his trainers and to alter his immediate environment." He can, upon being questioned about two differently shaped and differently colored objects, apparently understand which category (color or shape) is in question.

Such studies are extremely controversial, raising as many questions as they provide answers. Scientists adore wrestling with such questions—ranging from the exactitude with which the experiments are carried out (is there some subtle cuing going on by the human handlers which they aren't aware of?) to the ability of a nonhuman animal to use language in something like the way we do—such as referring to an object that is not present in time or space, the use of syntax, and so forth. If it is agreed (and it isn't), that these experimental animals are using rudimentary human language, then the question arises: Why don't they use it in nature? Or do they, and we just haven't noticed? It also raises the question of where this excess capacity comes from. If it is not employed in daily life in nature, then why does it exist? It is usually assumed that evolution does not produce overcapacity—in the way that a car manufacturer can make automobiles that can accelerate from zero miles an hour to sixty in twelve seconds, even though such use is not only illegal but certainly not necessary for survival on the road. Such questions are not irrelevant to the question of how human speech originated.

No less a thinker on such matters than Charles Darwin opined in *The Descent of Man* that language owes its origin to imitation. The question, of course, is "the imitation of what?" since there was no human language around to imitate (no one in those days had the advantages enjoyed by Alex the parrot) when speech first came into being. Controversies about the origins of speech have long raged, and there is little if any agreement about the matter to this day. So heated were the discussions of this topic at one time that since 1866 the Linguistic Society of Paris has prohibited all such discussion at its meetings. Most such discussions have been undertaken by linguists and anthropologists, which is reasonable enough, theirs being the study of human affairs of one sort or another, and it has fallen to them to specify what is so wondrously unique about human speech. Indeed, so wondrously unique is it that it seems almost to have occurred without any antecedent, as if by divine intervention or an abrupt and unprecedented lurch in the path of evolution.

A good reason for the contentiousness involved in this question is that it really isn't a scientific question at all. There is no way that anyone can test any hypothesis put forward for its explanation, since the events that occurred that led to human speech are not likely to be reproduced in any other living animal populations, and those events left no direct fossil traces. One can only guess what happened. Daring to go in where French academicians fear to tread, however, we think it is reasonable to look at the matter again in the light of what we now know about the vocal communication of nonhuman animals. That is to say, we take it as an article of faith that human speech, however unique, had to have arisen from some antecedent(s), so once we have specified what is unique about human speech, we can look from the bottom up, as it were, and determine what elements could have been present that circumstances might have triggered into becoming so astonishing a talent. To do so means keeping a large number of balls in the air at once, some of which have already been discussed in previous chapters and some of which must be introduced for the first time into the juggling act.

▌▎▌▎▌▎▌

FIRST, THE APPARATUS

It may seem to be putting the cart before the horse to ask when human speech might have first appeared in something like its present form, but there is indirect archeological evidence that is suggestive, if we assume that relatively complex language permitted early human beings to organize their lives in ways that their predecessors could not. For example, arguments continue unabated over the place of Neanderthals in the scheme of human evolution: They appeared some 100,000 or more years ago, and seem in many ways to have been human or at least very humanlike. Efforts have been made to show that the arrangements of the small bones involved in their vocal tracts made it unlikely that they were capable of articulate speech, but such efforts have largely failed to be convincing. Arguments also persist about their fate: Were they simply eliminated by competition from what amounts to modern humans (our direct line) which arose sometime later, perhaps 40–50,000 years ago or more, or were they absorbed into the human gene pool? What is more important (or at least more apparent) is that their style of doing things eventually vanished. Excavations that illustrate millennium after millennium of Neanderthal habitation show a simple range of tools and a simple arrangement of society that persisted, year in and year out, unchanged for tens of thousands of years—a basically boring toolkit of stones used for hammering and breaking things (presumably bones) and scraping things (presumably hides). Throughout Neanderthal time and geography, archeologists find only simple hearths in caves where women, accompanied by children, went about simple chores, while men pursued their activities from separate vantage points near the cave entrances.

Then, around 40,000 years ago, there was a sudden surge in creativity shown in the nature of tools in use— more elaborate scrapers, more finely made implements of an astonishing range, bespeaking an array of more complex activity and a more complex society. These are the remnants of what can be called modern humans— Cro-Magnon man, as this new creature is often called, followed soon by such *familiar* human activities as art— elaborate paintings on cave walls. Clearly the human revolution had occurred, and we were off on the trajectory that has led us to our present role in the world. It is assumed that something very much like complex human speech must have been in use at this time of emergence, but probably not long before.

Archeologists and paleontologists point out, however, that as far as can be told, the anatomical equipment presumed necessary for human speech already existed: the large, bilateral brain, fully upright posture, along with the use of fire, tools, and other behavior that is at least hominid in character. What would become modern human beings were, in a sense, preadapted for speech. This is suggested by the fact that the great flowering of complex artifacts and social organization that occurred in Europe and the Middle East 40,000 years ago did not occur in Asia until some 20,000 years later, even though members of the same human species inhabited Asia as early as Europe. Whatever the external conditions were that triggered the rather sudden explosion of cultural complexity in Europe, they did not pertain elsewhere until later. Just what those external conditions were is speculative, and an educated guess will be ventured in due course. Both Neanderthals and modern Homo sapiens were essentially brainy creatures, with a lot of complex wiring. They had intellect.

And so do a lot of other animals, as we were at pains to point out in Chapter Five, discussing the proposal of Nicholas Humphrey about the origins of intellect. To summarize briefly, the mental capacity which we can call intellect evolved primarily as a social tool, and only indirectly as a means of coping with the practical matters of daily subsistence. It arose as a means of dealing with increasingly complex social affairs—such as the existence within a single band or population unit of several generations, each with its own special requirements. The making of tools and their use—such as that twig a chimpanzee fashions to extract termites from a mound—can be arrived at by sheer imitation; it is not a continuing creative act of invention by succeeding generations. There is ample ethnographic evidence that in many human societies, especially those at the level of hunter-and-gatherer, technological information is learned by observation rather than verbal transmission. On the other hand, once a society has reached a certain level of social complexity, new pressures would exist internally that would act to increase this complexity still further.

Humphrey's point, as far as man is concerned, is that human ancestors, upon moving into the savanna from the trees, found an environment where technical knowledge began to pay new dividends: In that environment, pressures to prolong the learning period of children to give them even better schooling would have created a social system of as yet unrivaled complexity and, with it, a new challenge to intellect. Social understanding and foresight—a social intelligence—developed in the context of coping with family or band problems of interpersonal relationships, of gaining a personal advantage without tearing the social fabric upon which one's continuing existence depends. The development of such an intellect has been seen to be the result of one or another "arms race," a gradual ratcheting up of the ability to influence one's band mates, and this same process has been seen to exist in other social animals, in particular primates where something like deception (and the vigilance against deception) or reciprocal altruism (and the vigilance against welshing on a "deal") may be present. In other words, there is a continuum of intellect in evolution.

At the same time, we noted in Chapter Four that the

Motivational-Structural rules that appear to govern most of the short-distance vocalizations of birds and mammals suggest that an increasingly complex society would bring forth an increasingly complex or subtle array of vocalizations, and this appears to be borne out by the facts. In other words, an increasing subtlety of intellect and a concomitant increase in subtlety of communication were already under way long before what would eventually become humanity reared up on its hind legs and began carrying things around in its newly freed-up hands, eventually realizing the benefit of hauling a particularly suitable rock (or whatever) from where it was found to where it could be best put to use: that is, what we would recognize as a well-developed practical foresight and memory. And this intellect developed in response to social needs rather than purely practical ones. It has lately become popular in scientific circles to call this process "Machiavellian." (It is too bad that, as with "altruism," scientists have chosen an unfortunate word. While Machiavelli's political philosophy does evoke the kind of utter self-interest that appears to be prevalent in most animal behavior, it also calls to mind a highly developed cynicism. No nonhuman animal is capable of cynicism, of course, and the use of such words, loaded as they are with anthropomorphic implication, only serve to obscure matters further.)

It seems reasonable in any event to say that a certain level of sophistication of intellect existed among some animals, and in particular among primates, before human speech or even what we might think of as proto-speech occurred. But there are other basic requirements for speech that had to exist as well—certain matters of neural wiring. One example: Human language is believed to be processed chiefly in the so-called dominant hemisphere of the human brain, which for most people is the left hemisphere. In humans, this is associated with handedness and most people are right-handed. Additionally, the left hemisphere of the brain "attends" chiefly to the right ear. And in the rapid processing of sound, such as speech, the right ear–left hemisphere organization has been found to be superior, to have an advantage. This is particularly true in the case of what are called stop consonants (*b, d, g, p, t, k*) and, as linguists have pointed out, it is precisely these stop consonants that, when interposed among vowel sounds, make it possible for a great many words to be spoken and differentiated in a short period of time—one of the chief acoustic characteristics of human language universally. The new insights into animal communication we have been discussing have been based on a recognition that the acoustical *form* of an utterance plays a crucial role, indeed is indistinguishable from its function, so the reader's ears should have pricked up by now. The point here is that language is associated with hemispheric specialization of the brain.

Hemispheric specialization of the brain, however, has recently been found not to be an exclusive characteristic of mankind. For example, in 1988, scientists at the California Institute of Technology found that certain monkeys called macaques exhibited brain lateralization: The left hemisphere was better at distinguishing among tilted lines while the right hemisphere was better at discriminating among faces. Such tasks are called cognitive processing, and these experiments suggest that a lateralization of cognitive processing appeared in primates *before* either handedness or language, thus turning on its head the common assumption that the development of language was the engine that drove the lateralization process. At the same time, other scientists discovered a right-ear advantage in Japanese macaques in their discrimination of their species-specific *coo* vocalization. Lateralization has also been found in baboons.

Again, one of the peculiarities of speech which we alluded to earlier is the distinction between the sound *pa* and *ba*. The consonant *p* is unvoiced (it can be done merely with the lips and a little puff of air from the mouth) whereas the *b* is voiced. The distinction between one and the other, in fact, is not exactly clear: It lies along a continuum. The human ear, however, is generally capable of noting when *p* has become *b,* and that has much to do with one's perception of the timing of the onset of the voiced part, it coming later near the pure *p* end of the continuum and earlier near the *b* end. It turns out that rhesus monkeys have been shown experimentally to make the same kind of auditory distinctions between *pa* and *ba*. They also, it has been shown, have the articulatory apparatus to make such "labial" stop consonants though they do not apparently make such sounds in their natural vocalizations. They do *not* have the ability to make either a *g* or a *k* sound (their vocal tract is simply the wrong shape) but experiments show that they can make auditory distinctions between the two. Why? No one knows. But, as we have seen, these findings are of especial interest when one considers the new light that has been thrown on the crucial role of the perceiver in animal communication.

▌▎▌▎▌▎▌

EXCESS CAPACITY

In thinking about this "unnecessary" ability to distinguish consonant sounds among monkeys, several theoretical or even philosophical considerations arise immediately. First of all, evolution did not intend that there be human speech: It did not plan to set up the apparatus for speech (and the hearing and comprehending of speech) any more than there was some preexisting plan for the fish's gill to become the mammal's inner ear. Each minor alteration, created by natural selection working on varying anatomy within a species and its operation in a contemporary environment, is presumed to have served some contemporaneous need—or at least not to pose a significant disadvantage to the creature in question. Form would change incrementally, serving function, until fortuitously some other function was served as well or, eventually, instead. But there is something in the evolutionary process that has been called "excessive construction." In a sense, it is a "reserve" capacity to tolerate more severe conditions than the minimum that is typically required of the parent population.

For example, an animal may be able to run much faster than it actually needs to. The bark of certain trees may be resistant to fire, calling for an excessive amount of

investment in bark constituents when you figure that
most trees of that species probably won't get caught in
a forest fire. Such excessive construction may permit
creatures to expand into new environments, or survive in
radically altered ones. When this happens, natural selection would act very quickly to favor those with the excessive construction over those members of the species
without it. But even in such a radical situation, the new
wrinkle, the variation that works in new circumstances, is
essentially conservative. It permits the creature to go on,
as much as possible, doing what it has always done. A
tree more heavily sheathed in fireproof bark survives the
forest fire and goes on being a tree just as before: It
doesn't become mobile. In any event, the old shibboleth
that nature isn't wasteful does not seem to be borne out.
There is doubtless excess capacity all around us that will
never be triggered into use because the circumstances
are simply not right. (One need only consider that group
of human beings commonly referred to as couch
potatoes, sitting watching an endless stream of TV sit-
coms, to get the point.)

In any event, the rhesus monkey has evolved with an
excess auditory capacity, in a sense *proto*adapting it to
listen to a kind of vocal communication it would never
hear from its own kind. Beyond that, another principle
underlying evolutionary biology is that if distantly related groups of animals share a trait, it is a good bet that
it is a case of what is called "homology," meaning that
it is a trait they all inherited from a common ancestor,
rather than having arrived at it separately by convergent
evolution. (An example of convergent evolution is to be

seen in the finches of the Galápagos Islands, where similar feeding habits among unrelated birds both on the islands and the mainland led convergently to similar beak shapes. On the other hand, the mammalian ear presumably evolved once and got passed along to all of them; this is true homology.) We can therefore conclude that along with an increasingly well-developed social intellect, there also arose an increasingly sophisticated neural and anatomical apparatus for distinguishing subtleties among vocalizations, and this is not surprising in the light of discoveries that primate vocalizations are in fact quite subtle and are probably heard as discrete sounds, not just gradations along a continuum. As we see with *pa* and *ba,* there is some point where it becomes hard for us to detect when one call crosses some identifiable threshold into another call and carries a different freight for the listener. There is even something akin to syntax in certain monkey calls: In the wild, for example, titi monkeys repeat calls to form phrases and combine these phrases into sequences. Playbacks of these different sequences elicit different behavior from the monkeys.

In short, the possibility is now emerging that many of the differences between human and nonhuman primate communication are ones of degree rather than kind. It is possible to say that if the earliest hominids—the creatures like Lucy, who lived a million or so years ago—were derived from creatures with *at least* monkeylike communicative capacities, then their vocal-auditory machinery was probably far more ready to take on a "primordial" speech function than has previously been supposed.

It should be pointed out, in this connection, that much of the findings about the neuroanatomy of monkeys has been the result of invasive, surgical techniques that are no longer considered permissable in the case of the apes. The questions being asked are simply not considered important enough to justify the means. Indeed such means are increasingly frowned upon in many circles for any animal, unless the questions being asked are fundamental to medical science, and even then there is criticism. But the apparent capacity for chimpanzees to use sign language and other systems in a manner that approaches the use of human language is suggestive of another case of excessive construction or what, from hindsight, we might call protoadaptation to language. Still, it should also be pointed out that while studies of ape-human communication may have yielded helpful insights into how human children, particularly brain-damaged human children, may acquire language, there is no reason to believe that the way children acquire language is how a species of hominid acquired it. Human children are predisposed directly to acquire a language—a complicated one at that. Presumably, the emergence of "protolanguage" was slow in coming, possibly over thousands of generations, though once it hit some critical plateau it probably developed in complexity at a very great rate.

LANGUAGE DESIGN AGAIN

In Chapter Two, we took a fairly long look at the sixteen characteristics of human language, as proposed by Charles Hockett, many but not all of which are shared by other animal communication systems. One of these was semanticity, a precise meaning for a given utterance, and we have looked fairly coldly at semanticity in the likes of vervet alarm calls. At best, one could say that there is a kind of protosemanticity there, but protosemanticity is ample for the purposes at hand, which is to speculate on how a range of talents and capabilities could have come together in some way to form speech. It is possible that there is room for some further invention in the alarm calls and other utterances of the vervets: Some evidence exists that they have come up with a relatively new call to signal the presence of a human hunter with a dog. But a crucial aspect of such calls is that they exist in a *closed* system. With all the potential gradations of one call into another, each being a slightly different response to slightly different circumstances, there is a finite number of calls for any system, and they typically are responses to a certain relatively narrow range of circumstances— the presence of food or danger, the absence of band mates, and so forth. Except for matters of subtle degree, each call is exclusive.

But as Hockett pointed out in an investigation of the origins of language that he and a colleague, Robert Ascher, performed in the 1960s, an essential step in the evolution from a call system to language is the development of an *open* or productive system of vocal sounds. That is, the ability to freely emit completely new utterances, ones that have never been said before, in order to respond to completely new situations or to communicate about objects or events that are not present. And in speech (and written language) this is accomplished by what is somewhat mysteriously labeled *duality of patterning,* meaning that the utterances of a language consist almost entirely of elementary signaling units (such as *sig* and *nal*) which have no meaning in themselves but take on meaning when combined.

Such a system makes possible a nearly infinite number of combinations of phonemes—phonological components—whereas a system of calls is severely limited by the number of different sounds one can invent and remember. Think, in this sense, of the pictographic nature of early Chinese writing, which had virtually no relation to the sounds of vernacular words. Instead, by way of providing a common written language for a host of vernacular traditions of spoken language, individual characters stood for specific words. Eventually there were so many characters that only superscholars could remember them all and a process of simplification began, whereby abstract rather than pictographic units began to be used in a combinatory fashion. This has never been simplified to the point of the alphabet we use, in which twenty-six characters can take care of most ordinary requirements.

Any such system embodying this duality of patterning must be handed down culturally—that is, by each generation learning the entire system wholly anew. A call system can and is handed down genetically: Only in the rarest of circumstances is there anything in a call system that is invented anew; the degree of learning possible in a new generation is very narrowly proscribed by genetic capacity.

To bridge this dichotomy, Hockett and Ascher proposed that at some point in hominid evolution, at some point after our ancestors had been forced out of the trees and onto the savanna, where they confronted a largely new and more complex environment of food possibilities and dangers, they began to blend calls. For example, encountering food and danger simultaneously, they might have begun combining the two relevant calls into a more complex signal. If they began with a repertoire of, say, ten calls in their closed system, and began blending each with the others, they would before too long (or after a very long time, in fact) wind up with a repertoire of about 100 calls. But still a closed system. On the other hand, each blended call would consist of two previous calls, establishing a habit of building composite signals out of constituent parts.

From such a habit, Hockett and Ascher suggested, an open system would arise in which the constituents could become more and more abstracted until a language with duality of patterning emerged. Concomitant with this emergence would be the placing of a premium on the capacity for learning—and teaching—and thus on the genetic basis for that capacity. Hockett and Ascher went on to suggest that much of the learning of this new open system of vocalizations would be carried on away from the contexts in which the utterances themselves were directly relevant. In other words, the phenomenon of displacement would be encouraged. This, they said, might grow out of mammalian play.

Young mammals will often go through some of the preliminary motions of fighting when fighting is not what they apparently have in mind. One can see verbal play being added to physical play, with young animals taking up the fun of babbling on about things that aren't there. (It is worth pointing out in passing that adults today often learn a lot of new words, for better or worse, from their kids: Consider the grandfather who is doomed to go through the remainder of his life known to *everybody* as "Gampy," or worse.) With each incremental step along the way toward an open system of communication, the traditional activities of the band would be more aptly carried out. That is, its social affairs would be better managed so as to better manage, in turn, the practical matters of survival. A longer period of childhood dependency would have a racheting effect on the complexity of society, which would call forth more complexity in communication, and so on, until what we think of as language emerged. And once a certain level of complexity in communication was achieved, the drive to further complexity would be nearly exponential. Language ability would have become a paramount criterion for leadership and status within a band of humans, as it usually is among hunting-and-gathering societies we know of today. And there is suggestive evidence that, in such societies, the leaders tend to have more children that survive to maturity and reproduce.

As anthropologist Robbins Burling pointed out, in discussing the arms race leading to greater and greater linguistic capacity, the fact that some apes (among other animals) can learn to understand the rudiments of spoken language suggests that when one or a few early hominids began to use the rudiments of protolanguage and, later, true language, there is no reason to think they would not have been understood by their fellows, even if their fellows were not capable themselves of making such utterances. "All of us," Burlings says, "can understand words and all of us can understand dialects that we cannot reproduce. We are able to appreciate and be persuaded by those who speak more skillfully than we do ourselves. Nonpoets can be moved by poets."

■I■I■I■

VOILÀ!

So there we have it. Given the likely reproductive success of the "fast-talkers" in a band, it would not be too many generations before the band consisted of people who all shared the latest plateau of an open communications system or protolanguage, eventually leading to bands with what we would call language. And it would seem that this particular flowering took place only among modern humans, beginning some 40,000 years ago but occurring at different times in different places. The selective advantage of being a human with language would seem to be fairly obvious. Out of these selective pressures favoring those with an emerging language would have come the universal human trait, genetically dictated, of grammar, arising from more elemental syntactical usage.

Dialects would inevitably occur, groups of humans being separated from one another for at least most of the year if not for years at a time, and no one who pays any heed to the kaleidoscopic changes in teenage slang these days needs to be reminded of how fast a language can change. A question arises: What good is it, from an evolutionary standpoint, to have dialects? It has often been suggested that human dialects would have served as identity markers of a sort, tending to keep groups of people together, the ties that bind, and so forth. The apparently universal phenomenon among humans—the difficulty of learning a new language after puberty—would on the surface seem to bear this out. But there is an alternate view, one that is perhaps more plausible biologically.

In many primate societies if not all, some junior members have to leave the band. In certain marmosets it is daughters of reproductive age who must head out and shift allegiance to another band; among other primate groups, it is the males who disperse. This of course serves to cut down on inbreeding, the silent stalker of small populations, producing lowered reproductive fitness and various congenital afflictions. It is not at all unreasonable to imagine that human youths, equipped with a language capacity and their own dialect, kicked out of their own band by one means or another at just about the time they were of reproductive age and thus less able to master another dialect perfectly, would arrive in another band's purview capable of understanding its dialect and also of making themselves understood, but with an accent that made it clear they were unrelated: fresh genes, as it were. Such youths might have been welcomed, especially if they were graced with a superior linguistic ability and could sweet-talk their way into the new band's good graces.

Such an overall scenario for the development of language is plausible, if relatively vague, and speaks to the ultimate considerations involved—the longterm advantages. But it says little about the proximate causes, the immediate advantages of the incremental steps, and, more important, how those incremental steps took place. What were they and where did *they* come from?

THE CRUCIAL CIRCUMSTANCE

What is truly startling is that whatever the crucial steps were, they apparently took place only once. This is startling because there is a great deal of evolutionary convergence in so much of animal communication. The Motivational/Structural rules are perhaps the prime example of this convergence: They have come to apply to all known species of birds and mammals, including us. A moment's reflection shows that a great deal of human vocal communication obeys the M/S rules. We speak in a high register to babies. We typically speak in low, raspy tones when we are making threats. Much of the meaning of our words is conveyed by our tone of voice. Don't virtually all human voices rise toward the end of a sentence that is a question?

This legacy from our reptilian ancestors is a powerful and omnipresent one. And, at the same time, among a growing array of fellow primates, we are finding at least the auditory capacity to make the kinds of distinctions one needs to to understand human speech, along with the lateralization of the brain that is so deeply involved in our language capacity. It is safe to say that something like this kind of wiring existed in whatever protoape branched off several million years ago and became the progenitor of humanoids, which in turn led to hominids (like Lucy) and eventually to us.

Human paleontology is a bramble patch of argumentation, but it is highly likely that there were several hominid species wandering about at the same periods of time, all but one of which eventually became extinct. It is possible that some of them developed a highly sophisticated system of calls, maybe a system nudging toward an open system of the sort called for by Hockett. But it seems they never made it. For there is only one extant open system of language—human.

The selective pressures on the other hominid species were simply not sufficient, or the neural equipment of the other hominids was not succicient (or both), to call forth the crucial steps. Something or some things unique to humans among all the hominids, and unique to the human experience, called forth language. All the hominids had intellect to one degree or another, a certain amount of auditory capacity and other wiring, a complex society, and a variety of tools, and they managed quite well for unimaginably long periods of time without the benefit of true speech.

There is an anecdotal account that British zoogoers will typically shield their children's eyes when, say, the mandrills in the zoo begin to copulate, bustling the kids off to some other exhibit, while French zoogoers will point out such behavior to their children and even applaud. Sex seems rampant among animals, even in happy zoo animals, unfettered by the kinds of mores that we humans adduce to the procedure. But of course sex among animals is serious business (there appear to be no innocent flirtations, for example) and it is by no means unfettered or unconstrained. One of the chief constraints in mammalian sex across the board, with one exception, is that females are not always prepared to engage in it. Estrus may be seasonal, occurring once a year or even less frequently, or more often than annually. It is always obvious in one way or another, usually by scent to which mammals are well attuned, but there are numerous other signals, many of them visible, such as the swelling and coloration of the sex organs, or behavioral, including the vocal (as we saw among elephants). Owners of unspayed female dogs know how impossible it is to disguise estrus from those most interested. But it is not so with humans. Human females ovulate on a highly regular basis that can be calculated with considerable accuracy from the onset of menstruation (which is—or one assumes was in early bands of humans—quite noticeable.) On the other hand, it is not altogether likely that males among early human bands were bent on making such calculations.

The point is that estrus in human females is basically hidden and, for all poor old Grog knew, was probably unpredictable to the point of sheer mystery, if indeed old Grog was able to make the association between copulation and childbirth. It is probable that prehumans operated in this sphere without much ratiocination, rather like other primates. Perhaps even humans, for that matter: It seems that the earliest Greeks thought women could be impregnated by the wind or by a dip in the river—a frustrating time for men indeed, if they were concerned about the matter of paternity. And they probably were concerned on some level, since all other primates, as well as most other animals, including many birds, have evolved some means by which paternity is guarded. But if there is no way to tell when your mate is ready to copulate, meaning that she may be ready anytime, and in fact can and does copulate at any time through the year, then you, the male, have to be pretty protective about all this, making sure that you keep an eye on things at all times when she is not actively engaged in carrying a suckling infant around. Under these circumstances, you would exert all your wiles and charms and vigilance to see to it that she stayed close to you so that you could be sure her children were yours.

Meanwhile, from the female's perspective (which is probably a great deal more important in this scenario), thanks to the fact that you live in this complicated society where your children require the services of adults for an unconscionably long period of time before they can be considered competent, you will both be far better off if you can retain the continuing attention of their father, so that he will, as it were, help around the house (call it Mutual Assured Paternity). In primate terms, and probably even in hominid terms, this was a new requirement.

In a situation like this, sexual selection—that is, which male a female wants to be sire of her children—takes on a role as important or more so than mere natural selection—the ravages of nature upon one's body, the ability to find food, the succumbing to disease. The male-female bond becomes a crucial long-term matter and, once mating is complete, is nothing if not a social bond. It is not at all hard to imagine that the more subtly and capably the male, with his interests, and the female, with hers, could communicate with, manipulate, play on the expectations of, sell, persuade, wheedle the other into a more permanent relationship of mutual assistance and affection, the better off both would be in the matter of producing grandchildren, which is of course the name of the game.

THE FIRST STEP?

But that situation, however plausible from a zoological standpoint, still does not get to the possible first steps in "converting" a closed system of calls to an open system of at least protolanguage. It has been suggested that language itself might have arisen from gesture. After all, gestures are deliberate, one has volitional control over most of them. And volition seems also to be a cardinal feature of language. And living primates, especially apes, seem well able to employ gestures of one sort or another including such complex ones as those that make up sign language. But logic suggests that if one elaborated a communication system based on volitional gestures, at the same time carrying on nonvolitional vocal communication made up of calls, each being an automatic response to one or another of a handful of typical situations, then the conversion from gestures to spoken words would be a more distant leap than even a brainy hominid would have any need or impulse to achieve.

Merely from the standpoint of the conservatism inherent in evolution, it would seem best to find vocal antecedents for the first incremental steps from a closed system of calls, whines, groans, and growls, to an open, highly semantic system of words, syntax, sentences, and ultimately abstractions like ideas, and to the consciousness that one is communicating when one does so. What is basically unique about language is that it consists of phonemes, more or less arbitrary units that can be combined into meaningful noises we call words and sentences. And phonemes consist of vowels and consonants which are acoustically quite different items. They have different acoustical forms and it is the *combination* of vowels and consonants in the same vocalization that is fundamentally unique structurally to human language.

Try making up a dictionary out of nothing but vowels—*a,e,i,o,* and *u* and the various gradations you can work between them. You will wind up with a short dictionary and a bunch of sound-alike words, and it would take a lot of pausing between words to turn them into an interpretable sentence. But throw in some twenty consonants and the possibilities are virtually unlimited, particularly when you get to the point of using sentences that consist of more than three or four words. And, so far as we know, no living creature other than the human being combines vowels and consonants in the same vocalization.

Animal calls are made up almost entirely of vowel sounds, this being the means by which they express their state of mind, given the set of circumstances they are confronting at that particular moment. It has been argued that some of the more advanced call systems of certain primates like vervets suggest an amount of volitional control over what is emitted, but these arguments are so far not particularly convincing. The calls of animals (and presumably those of our ancestral forms) are and were largely unvolitional, however subtly graded they might have been or are in order to express subtly different states of mind. But there comes a point in an increasingly complex society when the continuous expression of sheer emotion becomes unproductive. There comes a time to bank one's feelings, or perhaps disguise them slightly, in order to accomplish other goals. As we've seen, the goose, over a long period of evolution, has come to disguise its natural fears in a hostile situation by emitting a nonvocal (nonvowel) sound—a hiss. In essence, this is a nonemotional consonant.

It is not at all unreasonable to look back over what we have perceived about animal communication, the Motivational/Structural rules that are universal among birds and mammals, the crucial importance of the perceiver, the common expression of mental states by means of vocal calls, the capacity among many primates to make distinctions among consonant sounds which are fundamentally neutral in emotional content (though they, too, can take on meaning over time), and imagine that some hominid began to use nonvocal, neutral consonant type sounds to dampen and disguise the emotional content of her calls, the better to play on the expectations of an interlocutor and accomplish some end which was of immediate advantage. A sufficient series of short-term advantages would add up to a long-term advantage and this, over generations, would prevail and probably at a suddenly accelerating rate.

And what might have led to this astonishing juxtposition of vowels and consonants? A perfectly plausible candidate is onomatopoeia, the imitation of other sounds. Suppose you were out collecting nuts and berries on the edge of the savanna and your child was struck by lightning and you felt the need to explain this calamity to your mate but you didn't have what we think of as words. You might try to pantomime the whole thing, ending up with a dramatic gesture mimicking the erratic pattern of a bolt of lightning through the sky. And your charade might go by without being understood. But if you were to add to the final gesture the explosive sound CRACK!, you might make yourself clearer. And old Grog would understand that it was lightning and not your irresponsible behavior that had wiped out his heir and, in time, the family would settle down again.

There is a small flycatcher which we in North America know by the name "pewee." This name is a pretty good representation of its most characteristic vocalization. But a spectrograph of a person saying *pewee* and the bird saying *pewee* will show that the bird doesn't pronounce the *p* as we do. We just hear it that way, being disposed to hearing consonants. But early humans, trying to suggest the presence (somewhere) of nature's "items," might well have "named" them by reproducing as best they could the sounds they make. And such sounds would be totally understandable by other, alien bands of people. This is borne out by a recent study that used long-suffering college students as guinea pigs. The students were presented lists of the names for a variety of animals in the language of the Huambisa, a remote South American tribe whose language was utterly foreign to any of the students. The test was to see if the students could, just by making out the sound of the names phonetically, tell which applied to birds and which applied to animals like fish that do not vocalize. The students—some 100 of them—performed significantly better than chance in recognizing the bird names, achieving an accuracy of some 58 percent overall. Some of the birds' names were in fact quite precise renditions of their calls, others were not. Instead, their names began with a "high" vowel sound, such as *eee,* which of course students of the M/S rules know suggests a smaller size. Typically such names referred to small birds, and typically the students got these right at their better-than-chance rate. So, perhaps in several ways, we learned to talk from other animals that couldn't.

What Darwin said in full was this: "I cannot doubt that language owes its origins to the imitation and modification, aided by signs and gestures, of various natural sounds, the voices of other animals, and man's own instinctive cries."

And so it seems to be.

Our achievement of language—speech, talk, discourse, call it what you will—seems unique, and it is. The poetry of Shakespeare, the formulae by which the physicist plumbs the otherwise unimaginable origins of the universe, the elaboration of thought in a philosopher's tome, the staccato reportage of the sportscaster, the car salesman's oleaginous rap, the lovesong, the demagogue's speech, the four-year-old boy saying "pterodactyl"—the marvels of human speech are utterly unique to us. But just below the surface, and often right on the surface, naked as a bird's alarm call, can also be heard the cumulative processes that began when some primitive amphibian grunted in the open air. Sir Isaac Newton, commenting upon his extraordinary achievements in the synthesis of the knowledge of physics, modestly said that he had been able to see so far because he stood upon the shoulders of giants. And we can speak so well, and write and compute and all the rest, because we too stand upon such shoulders—the shoulders of other life forms, each one a grand experiment in the business of life and of communication, each as urgently engaged by the challenges of living as we are, even the smallest a giant in its own right.

Bibliography

Darwin, C. 1965. *The Expression of the Emotions in Man and Animals.* The University of Chicago Press, Chicago.

Dawkins, R., and Krebs, J. R. 1978. "Animal signals: information or manipulation?" Pp. 282–312, in *Behavioural Ecology, An Evolutionary Approach* (J. R. Krebs and N. B. Davies, eds.). Blackwell Scientific Publications, Oxford.

Fenton, M. B. 1985. *Communication in the Chiroptera.* Indiana University Press, Bloomington.

Gish, S. L., and Morton, E. S. 1982. "Structural adaptations to habitat acoustics in Carolina wren songs." *Zeitschrift für Tierpsychologie 56:* 74–84.

Hartshorne, C. 1973. *Born to Sing.* Indiana University Press, Bloomington.

Hinde, R. A. (ed.). 1969. *Bird Vocalizations.* Cambridge University Press, Cambridge.

Humphrey, N. K. 1976. "The social function of intellect." Pp. 303–318, in *Growing Points in Ethology* (P. P. G. Bateson and R. A. Hinde, eds.). Cambridge University Press, Cambridge.

Jürgens, U. 1979. "Vocalizations as an emotional indicator, a neuroethological study in the squirrel monkey." *Behaviour 69:* 88–117.

Lanyon, W. E., and Tavolga, W. N. (eds.). 1960. *Animal Sounds and Communication.* Publication No. 7, American Inst. of Biological Sciences, Washington, DC.

Mitchell, R. W., and Thompson, N. S. (eds.). 1986. *Deception: Perspectives on Human and Nonhuman Deceit.* State University of New York Press, Albany.

Morton, E. S. 1975. "Ecological sources of selection on avian sounds." *American Naturalist* 108: 17–34.

———, and Shalter, M. 1977. "Vocal response to predators in pair-bonded Carolina wrens." *The Condor* 79: 222–227.

Morton, E. S. 1977. "On the occurrence and significance of motivation-structural rules in some bird and mammal sounds." *American Naturalist* 111: 855–869.

———, 1982. "Grading, discreteness, redundancy, and motivation-structural rules." Pp. 183–212, in *Acoustic Communication in Birds* (D. Kroodsma and E. H. Miller, eds.). Academic Press, Inc., New York and London.

———, 1983. "Animal communication: What do animals say?" *American Biology Teacher* 45: 343–348.

———, 1986. "Predictions from the ranging hypothesis for the evolution of long distance signals in birds." *Behaviour* 99: 65–86.

———, 1987. "The effects of distance and isolation on song-type sharing in the Carolina wren." *The Wilson Bulletin* 99: 601–610.

———, Gish, S. L., and van der Voort, M. 1986. "On the learning of degraded and undegraded songs in the Carolina wren." *Animal Behaviour* 34:815–820.

———, and Young, K. 1986. "A previously undescribed method of song matching in a species with a single song 'type,' the Kentucky warbler." *Ethology* 73: 334–342.

Moynihan, M. 1985. *Communication and Noncommunication by Cephalopods.* Indiana University Press, Bloomington.

Owings, D. H., and Morton, E. S. "Communication is *not* information exchange." To appear in *Brain and Behavioral Sciences.*

———, and Virginia, R. A. 1978. "Alarm calls of California

ground squirrels, *(Spermophilus beecheyi)." Zeitschrift für Tierpsychologie* 46: 58–70.

Page, J., and Morton, E. S. 1989. *Lords of the Air: The Smithsonian Book of Birds.* Smithsonian Institution Press, Washington, DC.

Payne, K., and Payne, R. 1985. "Large scale changes over 19 years in songs of humpback whales in Bermuda." *Zeitschrift für Tierpsychologie* 68: 89–114.

Ryan, M. J. 1985. *The Túngara Frog, a Study in Sexual Selection and Communication.* The University of Chicago Press, Chicago and London.

Sebeok, T. A. (ed.). 1968. *Animal Communication, Techniques of Study and Results of Research.* Indiana University Press, Bloomington.

Shy, E., and Morton, E. S. 1986. "The role of distance, familiarity, and time of day in Carolina wren responses to conspecific songs." *Behavioral Ecology and Sociobiology* 19: 393–400.

———, 1977. *How Animals Communicate.* Indiana University Press, Bloomington.

Smith, W. J. 1977. *The Behavior of Communicating: An Ethological Approach.* Harvard University Press, Cambridge, MA, and London.

Tembrock, G. 1977. *Tierstimmenforschung.* A. Ziemsen Verlag, Wittenberg Lutherstadt.

Tsipoura, N., and Morton, E. S. 1988. "Song-type distribution in a population of Kentucky warblers." *The Wilson Bulletin* 100: 9–16.

Walther, F. R. 1984. *Communication and Expression in Hoofed Mammals.* Indiana University Press, Bloomington.

Wilson, E. O. 1975. *Sociobiology, the New Synthesis.* The Belknap Press of Harvard University Press, Cambridge, MA.

Zivin, G. (ed.). 1985. *The Development of Expressive Behavior.* Academic Press, Inc., New York and London.

Index

About the Authors

EUGENE S. MORTON received his Ph.D. from Yale University in evolutionary biology, studying the acoustics of tropical habitats as sources of natural selection on bird song. His research interests concern the evolution of vocal communication, frugivory in birds, the evolution of colonial breeding in birds, avian breeding systems, behavioral ecology of migrant birds in the tropical regions, tropical biology of birds and plants, and dragonflies. In addition to the work he has published in scientific journals and magazines, he coedited *Migrant Birds in the Neotropics: Ecology, Behavior, Distribution, and Conservation* with Allen Keast. He works at the Smithsonian Institution's National Zoological Park in Washington, D.C., and is an adjunct professor at the University of Maryland.

JAKE PAGE is the author of hundreds of magazine articles and a dozen books in the realm of the natural sciences, including *Zoo: The Modern Ark, Pastorale,* and *The Smithsonian's New Zoo,* as well as *Hopi,* with his wife, Susanne Stone Page. He was formerly editor of *Natural History,* publisher of the Natural History Press, science editor of *Smithsonian,* and founder of Smithsonian Books.

Together, Morton and Page wrote *Lords of the Air: The Smithsonian Book of Birds.*